Lecture Notes in Biomathematics

ctd. on inside back cover

Lecture Notes in Biomathematics

Managing Editor: S. Levin

84

A. Dress
A. von Haeseler (Eds.)

Trees and Hierarchical Structures

Proceedings of a Conference held at
Bielefeld, FRG, Oct. 5–9th, 1987

Springer-Verlag
Berlin Heidelberg GmbH

Mathematics Subject Classification (1980): 92-06; 92A10, 92A15

ISBN 978-3-540-52453-3   ISBN 978-3-662-10619-8 (eBook)
DOI 10.1007/978-3-662-10619-8

2146/3140-543210 – Printed on acid-free paper

Table of Contents

Introduction

The "raison d'être" of hierarchical clustering **theory** stems from one basic phenomenon: This is the notorious non-transitivity of similarity relations. In spite of the fact that very often two objects may be quite similar to a third without being that similar to each other, one still wants to classify objects according to their similarity. This should be achieved by grouping them into a hierarchy of non-overlapping clusters such that any two objects in one cluster appear to be more related to each other than they are to objects outside this cluster.

In everyday life, as well as in essentially every field of scientific investigation, there is an urge to reduce complexity by recognizing and establishing reasonable classification schemes. Unfortunately, this is counterbalanced by the experience of seemingly unavoidable deadlocks caused by the existence of sequences of objects, each comparatively similar to the next, but the last rather different from the first. While in everyday life this may spur endless discussions about the usefulness of the applied classification scheme, much scientific progress seems to result in (and sometimes can even be identified with) ever better schemes. In fact, a deeper understanding of a given field of scientific investigation generally leads to a clearer distinction between superficial and deep-rooted, "structural" similarity. This in turn gives rise to (or can even be identified with) a better proposal for structuring the family of objects in question or at least can help to resolve previously observed deadlocks. Typical for the latter kind of progress is whether the problem of grouping the whale with all aquatic animals or with all mammals can be resolved by a better appreciation of morphological data. On the other hand the application of Darwin's theory of evolution to the Linnean program of classifying all living species according to class, order, family and genus is typical for the first kind. Nevertheless, even the deepest understanding of the structure of some family of objects may, and in general will, produce non-transitive similarity relations which cannot be interpreted unambiguously in terms of a hierarchical system of non-overlapping clusters.

Of course, in many branches of science there is no immediate necessity to ask for such a hierarchical system of non-overlapping clusters. There is, for instance, no inherent contradiction in the fact that classifying crystals according to their symmetry group as well as according to their chemical composition produces overlapping clusters — crystals with a very similar chemical composition may have different symmetry groups, crystals with the same symmetry group may have very different chemical compositions. However, even in such a situation, when it comes to producing a list of all identified crystals it is necessary to agree as to what comes first, the symmetry group or the chemical composition. In a way, such an agreement resolves the problem of overlapping by transforming the collection of partly overlapping clusters into a (perhaps slightly artificial) hierarchical system of non-overlapping clusters.

Similarly, while the architecture of ensembles of (more or less) related concepts does not need to be represented in terms of hierarchies (cf. [GWW87], [W89]), many psychological studies seem to reveal that the architecture of many such ensembles can indeed be approximated to a highly satisfying degree by hierarchies. This reflects probably the simple fact that hierarchical storage of knowledge and information allows rather efficient (though presumably not very creative) information management.

The situation is different in biology: according to our understanding of biological evolution we have every reason to believe that the presently living species have developed from common precursors by an essentially treelike branching process. This is reflected in and therefore can be recognized from the hierarchical structure of the system of all recent species. Hence, in biology, classification by means of non-overlapping cluster systems is not only an efficient bookkeeping device, but an essential tool for understanding biological evolution. Accordingly, it is here where the largest variety of methods for reconstructing phylogenetic trees by means of hierarchical classification procedures and for assessing the confidence one may attribute to any such tree has been proposed, analyzed and tested. And it is also here where the lack of absolutely unambiguous and objective criteria for the choice of the "correct" clusters is felt most strongly.

Still, biologists have never shied away from proposing phylogenetic trees, based on their interpretation of observed similarities and dissimilarities. Moreover, as shown more than twenty years ago in the landmark paper by W. M. Fitch and E. Margoliash [FM67], the advances in molecular biology give rise to impressive new and rather abstract data structures which can easily be used as input for automatic classification procedures. At first, these new advances seemed to promise an easy and final solution to all problems in biological taxonomy, once enough biomolecular data had accumulated. Nowadays, it appears that the most important progress initiated by these new methods is

- a much better and a much deeper understanding of the many delicate points which need consideration once (semi-)automatic classification procedures are applied to biological data

and

- their extension to a much wider range of applicability, not accessible to classical and in particular not to morphological taxonomic procedures, as e.g. the very early phases of the evolution of life (cf. [EW81a], [EW81b], [WDE86], [EWD88], [ELT89]), intraspecies evolution and diversification of genes, or the evolution of viral diseases (cf. [SH87], [DSP88]).

The conference **Trees and Hierarchical Structures**, organized jointly by the *Forschungsschwerpunkt Mathematisierung der Universität Bielefeld* and the *Sektion Datenanalyse und Numerische Klassifikation der (westdeutschen) Gesellschaft für Klassifikation* in fall 1987, was devoted in particular to study and discuss the various **theoretical** problems relating to the application of hierarchical clustering procedures. We tried to bring together scientists from various branches of science, in particular from biology, psychology and mathematics, all of which shared a common interest in the development and/or the application of such procedures. Not all the material discussed during this conference is represented in these proceedings, mainly because it was about to be published elsewhere already (cf. for instance [Fel83], [WE86], [G88], [BD89], [GWW89],). Still, the papers collected in this volume give a very useful overview of the many facettes of the problems, involved in applying cluster analysis to real data sets.

The papers by P. O. Degens et al., by P. L. Williams and W. M. Fitch and by D. and P. Penny discuss particular aspects of the tree reconstruction process. The first paper assumes the data to be given in form of a distance matrix. The authors discuss a variety of statistical models and procedures for evaluating the accuracy and the stability of estimated phylogenies. The usefulness of their methods is exemplified by applying them to the DNA-DNA hybridization data of C. G. Sibley and J. E. Ahlquist ([SA85]) concerning *New World Suboscine Passerine Birds*. Their methods provide substantial statistical evidence for "the involved part of the classification suggested by Sibley and Ahlquist".

The papers by P. L. Williams and W. M. Fitch and by D. and P. Penny address the special sequential structure of most biomolecular data. P. L. Williams and W. M. Fitch discuss algorithmic aspects of one important step in the search for most parsimonious (or Steiner) trees in sequence space, namely the optimal realization of a given tree topology. While this problem has been solved relative to uniformly weighted substitutions many years ago (cf. [F71] and [H73]), the extension to non-uniformly weighted substitutions is much younger (cf. [SC83] and [WF88], for a more general approach relating this problem to the minimization problem for certain spin glass Hamiltonians see [D86]). Based on such algorithms the authors have also tried various ways to attach weights to individual positions of a given (aligned) sequence family by a dynamic weighting process. The basic idea is to distinguish unreliable "hot spots" from positions, where changes seem to be of phylogenetic significance. They report a number of promising and interesting results and experiences, obtained that way.

D. and P. Penny discuss ways to use parallel computing in the search for most parsimonious trees in sequence space to get around (at least for reasonably sized data sets) the combinatorial explosion created by the vast number of $1 \cdot 3 \cdot 5 \cdot \ldots \cdot (2n-5)$ (non-degenerate) tree topologies definable on an n-set. Their suggestions are very persuasive and many of them have already been (or soon will be) implemented successfully. We share their hope that the work referred to in their paper should "convince more people that the study of evolutionary trees should accept only the highest scientific standards" or, in other words, that it is possible (and should be obligatory) nowadays in such investigations to meet the "minimum requirements of a good study" which they summarize as follows: "For any set of sequences we need to be able to demonstrate objectively that a binary tree is a good representation of the data (as opposed for example to a network — a graph with cycles). It should be possible to suggest a labelled tree that is the best predictor, as well as to indicate the range of trees that could still be possible as more data becomes available. A complete study should include an indication of the mechanism of evolution that generated the sequences, for example, whether all sites were equally likely to change, whether rates of evolution had varied between lineages." [PP89]

The paper by H. Abdi gives a rather useful introduction in and an overview of the purpose, the history, and the uses of additive tree representations in various fields — with particular emphasis on when and how to use them in H. Abdi's own field of psychology.

The papers by H.-J. Bandelt and A. v. Haeseler and by H. Hoffmann and H. Kühn apply tree reconstruction methods to explicit biological data sets. H.-J. Bandelt and A. v. Haeseler discuss the phylogeny of Prochloron which may or may not be a modern counterpart of the ancestor of the green chloroplasts. Using the data presented by E. Seewaldt and E. Stackebrandt ([SS82]) they discuss some important, but delicate points concerning their interpretation in terms of minimization of various objective functions. In this way they try to meet the "minimum requirements of a good study" as outlined by D. and P. Penny. Their results support essentially the tree suggested by E. Seewaldt and E. Stackebrandt which in the meantime has also been corroborated by additional experimental evidence (cf. [TBW89], but see also [MG89] which contradicts these results). H. Hoffmann and K. Kühn discuss the intraspecies evolution and diversification of the collagen fibril gene substantiating the claim that abstract classification procedures can be used successfully in such a context, too.

The last two papers do not deal with problems in biology or psychology. I. Althöfer explains an algorithm for treelike recursion networks which can be used in chess programs and tries to estimate the best move in a given situation from estimates concerning moves further down the tree (that is, later on in the game). This problem and its solution contrast in an interesting way with the algorithm presented by P. L. Williams and W. M. Fitch.

Finally, the rather short note by U. Höhle demonstrates that there is much more what can be said from a purely mathematical and modeltheoretic point of view concerning the above mentioned notorious non-transitivity of similarity relations. The ideas, presented by U. Höhle, are somewhat related to the injective hull construction for metric spaces, considered in [Is64], which was rediscovered independently in [D84] in the very same context (that is, in the search for constructions which can substitute for additive tree reconstructions in case the data do not satisfy the conditions which are known to be necessary and sufficient for the existence of such a tree). These ideas served as a sound mathematical background for the ideas developed jointly with H.-J. Bandelt later on (cf. [BD89]). We believe that the still more general and more abstract ideas presented by U. Höhle can, if thoroughly digested by more applied minded people, serve the same purpose for a still much broader structural approach to the problems of hierarchical cluster analysis.

We are grateful to Martina Diegelmann for her tireless support during the course of the conference and for her help in preparing the present Proceedings volume.

Bielefeld, May 8, 1989

Andreas W. M. Dress

Arndt von Haeseler

References

[BD89]   Bandelt, H.-J., Dress, A. W. M.: Weak hierarchies associated with similar-
          ity measures — an additive clustering technique. Bull. Math. Biol.**51**, 133
          – 166 (1989).

[DSP88]   Dopazo, J., Sobrino, F., Palma, E. L., Domingo, E., Moya, A.: Gene en-
          coding capsid protein VP1 of foot-and-mouth disease virus: A quasispecies
          model of molecular evolution. Proc. Natl. Acad. Sci. USA **85**, 6811 –
          6815 (1988).

[D84]   Dress, A. W. M.: Trees, tight extensions of metric spaces, and the co-
          homological dimension of certain groups. Adv. in Math. **53**, 321 – 402
          (1984).

[D86]   Dress, A. W. M.: On the computational complexity of composite systems.
          In: Fluctuations and Stochastic Phenomena. L. Garrido (ed.), Springer
          Lecture Notes in Physics **268**, 377 – 388 (1986).

[ELT89]   Eigen, M., Lindemann, B., Tietze, M., Winkler-Oswatitsch, R., Dress, A.,
          v. Haeseler, A.: How old is the genetic code. Statistical geometry of tRNA
          provides an answer. to appear in Science.

[EWD88]   Eigen, M., Winkler-Oswatitsch, R., Dress, A.: Statistical geometry in se-
          quence space: A method of quantitative sequence analysis. Proc. Natl.
          Sci. USA **85**, 5913 – 5917 (1988).

[EW81a]   Eigen, M., Winkler-Oswatitsch, R.: Transfer-RNA: The early adaptor.
          Naturwissenschaften **68**, 217 – 228 (1981).

[EW81b]   Eigen, M., Winkler-Oswatitsch, R.: Transfer-RNA, an early gene? Natur-
          wissenschaften **68**, 282 – 292 (1981).

[F71]   Fitch, W. M.: Towards defining the course of evolution: minimum change
          for a specific tree topology. Syst. Zool. **20**, 406 – 416 (1971).

[Fel83]   Felsenstein, J.: Statistical inference of phylogenies. J. R. Statist. Soc. A.
          146, 246 – 272 (1983).

[FM67]   Fitch, W. M., Margoliash, E.: Construction of phylogenetic trees. Science
          155, 279 – 284 (1967).

[G88]   Ganter, B.: Composition and decomposition of data in formal concept
          analysis. In: Classification and Related Methods of Data Analysis. H. H.
          Bock (ed.). Elsevier Science Publisher, p. 561 – 566 (1988).

[GWW87] Ganter, B., Wille, R., Wolff, K. E. (eds.): Beiträge zur Begriffsanalyse, BI Wissenschaftsverlag, Mannheim, Wien, Zürich (1987).

[H73] Hartigan, E.: Minimum mutation fits to a given tree. Biometrics **29**, 53 – 69 (1973).

[Is64] Isbell, J. R.: Six theorems about metric spaces. Comment. Math. Helv. **39**, 65 – 74 (1964-1965).

[MG89] Mouden, C. W., Golden, S. S.: psbA genes indicate common ancestry of prochlorophytes and chloroplasts. Nature **337**, 382 – 285 (1989).

[PP89] Penny, D., Penny, P.: Search parallelism, comparisons and evalutation: Algorithms for evolutionary trees (this volume).

[SC83] Sankoff, D., Cedergren, R. J.: Simultaneous comparison of three or more sequences related by a tree. In: Time warps, string edits, and macromolecules. D. Sankoff & J. B. Kruskal (eds.), Addison-Wesley, London, 253 – 263 (1983).

[SA85] Sibley, C. G., Ahlquist, J. E.: Phylogeny and classification of New World Suboscine Passerine birds (Passeriformes: Oligomyodi: Tyrannides). American Ornithology Union Monograph **36**, p. 396 – 426 (1985).

[SS82] Seewaldt, E., Stackebrandt, E.: Partial sequences of 16S ribosomal RNA and the phylogeny of Prochloron. Nature **295**, 618 – 620 (1982).

[SH87] Steinhauer, D. A., Holland, J. J.: Rapid evolution of RNA viruses. Ann. Rev. Microbiol. **41**, 409 – 433 (1987).

[TBW89] Turner, S., Burger-Wiersma, T., Giovannoni, S. J., Mur, L. R., Pace, N. R.: The relationship of a prochlorophyte Prochlorothrix hollandica to green chloroplasts. Nature **337**, 380 – 382 (1989).

[W89] Wille, R.: Lattices in data analysis: How to draw them with a computer. In: J. Rival (ed.), Algorithms and Order, Dordrecht-Boston, Reidel, p. 33 – 58 (1989).

[WDE86] Winkler-Oswatitsch, R., Dress, A., Eigen, M.: Comparative sequence analysis. Chemica Scripta **26B**, 59 – 66 (1986).

[WE86] Wolters, J., Erdmann, V. A.: Cladistic analysis of 5SrRNA and 16SrRNA secondary and primary structure — the evolution of eukaryotes and their relation to archaebacteria. J. Mol. Evol. **24**, 152 – 166 (1986).

[WF88] Williams, P. L., Fitch, W. M.: Weighted parsimony: when not all changes have the same value. Submitted to Mol. Biol. and Evol. (1988).

RECONSTRUCTION OF PHYLOGENIES BY DISTANCE DATA: MATHEMATICAL FRAMEWORK AND STATISTICAL ANALYSIS

Paul O. Degens, Berthold Lausen, Werner Vach

Department of Statistics, Dortmund University
Postfach 500 500 , D4600 Dortmund 50

1. Introduction
2. Mathematical Setting
3. Stochastic Models
4. Maximum Likelihood Estimation
5. Approximation by Hierarchical Clustering
6. Judgement and Comparison of Hierarchical Clustering Methods
7. Accuracy and Stability of Estimated Phylogenies
8. Model Verification
9. Example: New World Suboscine Passerine Birds

1. Introduction

The reconstruction of phylogenies is an important problem in biological systematics since Darwin recognized natural selection as scientific cause of biological evolution. Since the early beginnings, cp. Haeckel's rule of 'recapitulation of phylogeny in ontogeny', a main problem in the reconstruction of phylogenies has been to measure homology, i.e. the natural similarity of species given by common ancestors and not a similarity given by adaptation to a common environment.

This measuring was first done qualitatively, later on also by measuring lengths and weights of less or more important features. Hennig's proposals give a good line for the reconstruction of phylogenies, if we know plenty of high-specialized structures in a set of species. But assuming a restricted knowledge about the members of a group of species, mathematical analysis of the data may give some evidence. Most methods in numerical taxonomy give proposals for phylogenies, but without any hint about the accuracy of these proposals.

Today there is growing interest in biochemical systematics, based on genetical (dis-)similarity or sequence data, cp. Ferguson (1980). Especially DNA-DNA hybridization data provide excellent information about a distance in DNA evolution. Our analysis and proposals cover this kind of dissimilarity data, as we believe that DNA-DNA hybridization data is - because of its overall character - a valuable tool in the analysis of DNA evolution and for the reconstruction of phylogenies. Today the statistical analysis of sequence data is limited to a little part of the genetic material, a certain protein for example. For example Felsenstein (1983) and Weir (1988) discussed the statistical analysis and modeling of sequence data.

Methods known as hierarchical cluster methods are used for the analysis of dissimilarity matrices produced by DNA-DNA hybridization (for a general view of hierarchical cluster analysis see e.g. Bock, 1974, Jardine & Sibson, 1971 or Sneath & Sokal, 1973). Especially the agglomerative average linkage method (UPGMA) is common. Mathematical and statistical characterizations of well known agglomerative methods were given by Degens (e.g. 1983a,b,1985,1988), cp. section 4 and 5. Using the simple additive error model proposed by Degens (1983b), Ostermann & Degens (1984,1986) performed a Monte Carlo study for the UPGMA and analysed the probability to detect certain true clusters as a characteristic of the method, cp. also Lausen & Degens (1987), Vach & Degens (1988a,b,c) and section 3 and 6. Different error distributions are analysed. Vach & Degens (1986,1988b) consider possibilities to construct more robust agglomerative methods.

There are a lot of different algorithms to do hierarchical classification by dissimilarity data. But there is often a lack of mathematical and statistical characterizations regarding applied problems. Especially in most cases we do not know anything about the accuracy or inaccuracy of the estimator, i.e. of the method giving provisional reconstructions of phylogenies. Therefore our paper presents and reviews a mathematical and statistical approach which solves the problem above to some extent.

2. Mathematical Setting

We recall the two well known mathematical settings to describe phylogenies, the dendrogram (ultrametric) and the additive tree , as follows:

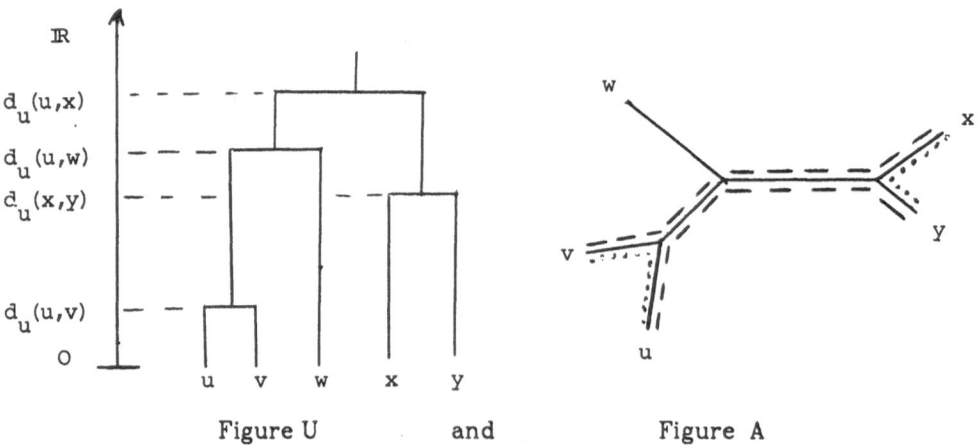

Figure U   and   Figure A

An additive tree may be defined as a graph theoretical tree, where every edge is associated with a positive (or nonnegative) real number. The sum of the values of the egdes connecting two exterior vertices defines a distance, which is called an *additive tree metric*, cp. figure A.

The existence of a root is necessary for the interpretation of an additive tree as a dendrogram. The sum of the numbers associated with the edges between the root and any exterior vertex is called its distance from the root. If an interior point in an additive tree can be defined as the root and if the distance from the root to all exterior vertices is constant, the tree is a dendrogram. Additive tree metrics, defined by these dendrograms, are called *ultrametrics*, cp. figure U.

For a formal description of ultrametrics and additive tree metrics let S be a non empty finite set of operational taxonomic units - e.g. species, strains of bacteria etc. - called points or objects. A symmetric map $d:S \times S \rightarrow \mathbb{R}$ with $d(x,x)=0$ for all $x \in S$ is called a distance on S. A distance d_u is an ultrametric, if the ultrametric inequality holds:

$$d_u(x,y) \leq \max(d_u(x,z), d_u(z,y)) \quad \forall \; x,y,z \in S. \tag{U}$$

A distance is an additive tree metric, if the additive tree inequality holds:

$$d_a(x,y) + d_a(u,v) \leq \max(d_a(x,u) + d_a(y,v), d_a(x,v) + d_a(y,u)) \; \forall \; x,y,u,v \in S. \tag{A}$$

There are useful equivalent formulations for (A):

Four points condition : Any four points of S can be called x, y, u, v, so that

$$d_a(x,y) + d_a(u,v) \leq d_a(x,u) + d_a(y,v) = d_a(x,v) + d_a(y,u), \tag{A'}$$

and for (U):

Any three points of S can be called x, y, z, so that

$$d_u(x,y) \leq d_u(x,z) = d_u(y,z). \tag{U'}$$

In (A') and (U') - and of course also in (U) and (A) - it is necessary to allow for coincidences of some of the points.

The equivalence of ultrametrics and dendrograms was shown by Johnson (1967), Hartigan (1967) and Jardine, Jardine & Sibson (1967), the equivalence of additive tree metrics and nonnegative weighted trees by Buneman (1971), Dobson (1974) and Simoes Pereira (1969). It is easily seen , that (U') implies (A'), hence ultrametrics are special additive tree metrics.

Furthermore ultrametrics can be described equivalently by isotonic functions on hierarchies (cp. Bock, 1974). A class H of subsets of S provided with inclusion as natural ordering is called a *hierarchy* (on S), iff the following three conditions hold:

(H1) $S \in H, \phi \notin H$,

(H2) $A, B \in H$ implies $A \cap B \in \{\phi, A, B\}$,

(H3) $\cup \{A \in H \mid A \text{ minimal in } H\} = S$.

Each ultrametric d_u induces a hierarchy

$$H(d_u) := \{\{x \in S \mid d_u(x,y) \le \varepsilon\} \mid \varepsilon \in \mathbb{R}_0^+, y \in S\}$$

and an isotonic function $d_u^*: H(d_u) \to \mathbb{R}_0^+$ defined by $d_u^*(H) := \max\{d_u(x,y) \mid x,y \in H\}$. d_u^* can be defined analogously on each hierarchy $H' \supseteq H(d_u)$ and $d_u^*(H) = 0$ holds for any minimal H in H'. The definite extension $H_0(d_u) := H(d_u) \cup \{\{x\} \mid x \in S\}$ of $H(d_u)$ is of special interest. Conversely an isotonic function $d_u^*: H \to \mathbb{R}_0^+$ on a hierarchy H with $d_u^*(H) = 0$ for all minimal H in H induces an ultrametric d_u defined by $d_u(x,y) := \min\{d_u^*(H) \mid x,y \in H \in H\}$, and $H(d_u) \subseteq H$ holds.

The definition of lower neighbours provides us with a valuable tool to handle hierarchies: $A \in H$ is called a *lower neighbour* of $B \in H$, iff $A \subsetneq B$ and there is no $C \in H$ with $A \subsetneq C \subsetneq B$. By $N(C,H) := \{A \in H \mid A \text{ lower neighbour of } C \text{ in } H\}$ we denote the set of lower neighbours of $C \in H$. $N(C,H)$ is empty, iff C is minimal; otherwise $N(C,H)$ is a partition of C.

Ultrametric or Additive Tree Models and the Molecular Clock

Describing phylogenies by dendrograms or additive trees is an idealization like the concept of mass points in physics. But taking this into account the true phylogenetic history is a dendrogram or an ultrametric in real time. So we suggest to fit an ultrametric model (with additive error for simplicity) first. Afterwards lack of fit of the estimated ultrametric could be a hint for the necessity of a more complicated model, i.e. the error may result from different components. Modifying the error structure it may happen that there are no good estimators for the underlying ultrametric. Then it may be appropriate for statistical reasons to estimate a generalization of an ultrametric; e.g. an additive tree metric.

A more technical argument for this procedure is, that the statistical and mathematical problems caused by the set of ultrametric trees are less complicated than the problems by the additive trees in general. Treating the reconstruction of the phylogeny of n species as an estimation problem, we must estimate $n - 1$ parameters for the ultrametric, but $2n - 3$ parameters for the additive tree (disregarding the estimation of the topology of the tree). A working rule derived from this relationship is: we may have a gain in accuracy in estimating additive trees, only if the violation or lack of fit of the ultrametric model is relatively high. We suggest a measure of lack of fit in section 8.

The statement ' ... the uniform average rate (UAR) is nothing more (or less) than the inevitable statistical result of averaging over billions of nucleotids and millions of years' (Sibley & Ahlquist, 1984) is reasonable, but it is not generally accepted. Arguments in favour of an (ultrametric) dendrogram are :

- the true phylogenetic history is ultrametric in real time,
- for the true phylogeny there exists a root,
- estimation of additive tree models causes more statistical and algorithmic problems,
- we must estimate for an additive tree about twice the number of parameters of an ultrametric tree (disregarding the topology),
- the gain of flexibility does not justify the loss of data information in an additive tree model, if the number of species is limited,
- a less complicated modeling is essentially proposed by the general scientific maximum parsimonium principle.

Having estimated the phylogeny as an (ultrametric) dendrogram, the obtained residuals from the data may allow the assumption of the uniform average rate (UAR) of DNA mutations or for short of a molecular clock , or at least may allow to work with such an assumption. Our proposal is to assume any form of an UAR, if we are not forced by some evidence from the data to reject the ultrametric model.

3. Stochastic model

In the literature there is a discussion about probabilistic models for statistical methods regarding the estimation of phylogenies. Compare the arguments of Farris (1985) and Felsenstein (1986). Some years ago there was a similar discussion in glottochronology, Dobson et al. (1972) give a clear description of the statistical point of view. We believe that such models, even if they are not quite accurate, may help to improve both the biological and statistical analysis. A method of measurement can be improved only if there is some insight in its inherent inaccuracies and in the lack of fit of the underlying statistical models. The analysis of these inaccuracies, e.g. by a residual analysis, may improve modeling and help to improve measurement methods, too, cp. section 8.

In order to evaluate provisional reconstructions of phylogenies we generalize the simple measurement error model of Degens (1983a,b), also cp. Baker (1974).

$$d = d_u + e \qquad or \qquad (3.1)$$

$$d(x,y) = d_u(x,y) + e(x,y) \qquad \{x,y\} \subset S$$

with S set of $\#(S) = n$ different objects (species, strains, OTUs),

    $(x,y) = (y,x)$ pairs of different objects without an order

        [abbreviation of ($\{x,y\}$)],

    $d(x,y)$ observable distance between species x and y,

    $d_u(x,y)$ true ultrametric distance,

    $e(x,y)$ random noise, error or fluctuation.

In treating the reconstruction of the phylogeny of n species as an estimation problem, it is difficult to find an estimator with good statistical properties, such as unbiasedness, minimum mean square error, good asymptotics for a growing number of species, and robustness. For mathematical reasons there exists no unbiased estimator in appropriate models and it is a difficult problem to find good robust estimators. The combined structure of an ultrametric, given by the topology of the dendrogram and the time levels, makes a rigorous statistical analysis very hard.

Degens (1983b) assumed e as independent identically distributed (i.i.d.) for trivial simplification. Then it is possible to derive the single linkage method as maximum likelihood estimator (MLE), if the distribution of e has a decreasing density on $[0,\infty)$. The complete linkage method can be seen as a MLE for a distribution with increasing density on $(-\infty,0]$. Similar results are derived for other distributions, cp. Degens (1983,1985), especially for the normal distribution and the average linkage method, cp. section 4 and 5.

Improvements of the stochastic model (3.1) are possible, taking into account the various causes of variability of the measurements and of the lack of fit of the ultrametric model. The experimental error resulting from the technique of DNA-DNA hybridization and of measuring the degree of dissimilarity may consist of components depending e.g. on

    each repetition of the measurement,

    each pair (x,y) of species,

    each species   or

    each true value, which is measured (heteroscedastic error).

Of course these components are chosen to meet both biological and statistical requests. More exact and appropriate components may be identified by a rigorous analysis of the hybridization and measuring method, cp. Lausen & Degens (1986,1988) and Felsenstein (1988). Another source of variability is given by the (random process of real) evolution, the random fluctuations of e.g. the DNA evolution.

For simplification and in order to make the analysis feasible it is senseful to assume independence of the error components of the measurement and to neglect heteroscedastic models. If the most important error components are identified, it may be possible in some instances to reduce the heteroscedastic influence by an appropriate transformation. But random fluctuations of evolution depend on the evolutionary process, giving a special tree-like covariance structure in a Gaussian branching process: The covariance $Cov(d(x,u),d(y,v))$ is proportional to the length of common evolution between x and u on one side and y and v on the other side.

Example:

$$Cov(d(x,y),d(u,v)) = 0$$

No common evolution.

$$Cov(d(x,u),d(y,v)) = c\,[(d_u(x,u)-d_u(x,y))+(d_u(u,x)-d_u(u,v))]$$

The first difference corresponds to the common evolution of the last common predecessor of x and y from the root, the second difference to the common evolution of the predecessors of u and v from the root. c denotes a constant. (This situation is marked in the above figure.)

$$Var(d(x,u)) = 2c\,d_u(x,u)$$

The same case with $x=y$ and $u=v$.

$$Cov(d(x,u),d(y,z)) = c\,(d_u(x,z)-d_u(x,y))$$

Common evolution between the predecessors of x and y and the splitting of $\{x,y,z\}$ in $\{x,y\}$ and $\{z\}$.

$$Cov(d(x,u),d(x,z)) = c\,d_u(x,z)$$

Common evolution of x from the latest common predecessor of x and z.

Hence we arrive at the following model with covariance structure (cp. Lausen & Degens, 1986, 1988):

$$d(x,y,k) = d_u(x,y) + \sum_{\nu \in P(x,y)} l(\nu) + e_0(x,y) + e_1(x,y,k), \qquad (3.2)$$

with legend of (3.1) and $k=1,...,r(x,y)$ number of replicates,

$d(x,y,k)$ observable distance between species x and y (k-th replication),

$P(x,y)$ set of edges ν on the path from x to y in the dendrogram seen as tree,

$l(\nu)$ random fluctuations of the evolution,

$e_0(x,y)$ error component of each pair of species, the same in each repetition,

$e_1(x,y,k)$ random inaccuracy of a single measurement,

l, e_0, e_1 independent vectors with independent Gaussian components,

$E(e_0(x,y)) = E(e_1(x,y,k)) = 0$,

$Var(e_0(x,y)) = \sigma_0^2 \geq 0$, $Var(e_1(x,y,k)) = \sigma_1^2 > 0$,

$E(l(\nu(C,A))) = 0$, $Var(l(\nu(C,A))) = \sigma_\gamma^2 (d_u(C) - d_u(A))$, $\sigma_\gamma^2 \geq 0$,

where $\nu(C,A)$ is the branch between a cluster C and a lower neighbour A of C,

$d_u(C)$ and $d_u(A)$ are the ultrametric cluster levels of C and A.

Note: e_1 represents a typical measurement error, varying, but i.i.d. for all repetitions, whereas e_0 is a typical block effect.

$d(x,y,k)$ has a tree-like covariance structure for $\sigma_\gamma^2 > 0$. $d_u(x,y) + \sum l(\nu)$ is roughly an additive tree metric with ultrametric expectation, but allowing negative branch lengths.

If the error depends on the structure (covariance structure), which has to be estimated, too, or on the estimated cluster levels (heteroscedastic models) a rigorous statistical analysis may be very hard, the formulae are complicated and interesting values are given only in an implicit way. For simplification we mostly neglect the special error component resulting in the covariance structure in this paper, but try to justify this by the data. This error component may be small in the evolution of younger taxa, but may be of large influence for old taxa, especially for estimation of split points of archaebacteria and eubacteria. We also avoid heteroscedastic error components. Then model (3.2) degenerates for $\sigma_\gamma^2 \to 0$ to the model

$$d(x,y,k) = d_u(x,y) + e_0(x,y) + e_1(x,y,k). \qquad (3.3)$$

In the Gaussian model (3.3) without covariance structure replications of measurements with fixed accuracy can for mathematical reasons simply be amalgamated by averaging, giving smaller variances. Assuming that the replications of measurements are of identical accuracy, we get with

$$\bar{d}(x,y) = \frac{1}{r(x,y)} \sum_{k=1}^{r(x,y)} d(x,y,k) \qquad \text{and}$$

$$Var(\bar{d}(x,y)) = \frac{\sigma_1^2}{r(x,y)} + \sigma_0^2 = \sigma^2(x,y)$$

the error model:

$$\bar{d}(x,y) = d_u(x,y) + e(x,y) \tag{3.4}$$

with independent Gaussian $e(x,y)$ for $\{x,y\} \subset S$ with $E(e(x,y))=0$ and $Var(e(x,y)) = \sigma^2(x,y) \in \mathbb{R}^+$ fixed, known. The case $\sigma^2(x,y) = \infty$ corresponds sensefully to $r(x,y)=0$, i.e. $d(x,y)$ is not measured, but a missing value. The models (3.2), (3.3) and (3.4) are called *balanced*, if the number of replicates is constant for all pairs of species.

4. Maximum Likelihood Estimation.

By the maximum likelihood (ML) principle one can look for an appropriate estimator of d_u in the model (3.4), cp. Degens (1983,1988). The following notation is useful for the ML-estimation and for the generalization of average linkage methods - especially for the agglomerative one - to weighted distances and missing values:

$$w(x,y):=w(y,x):=\frac{1}{\sigma^2(x,y)} \in \mathbb{R}_0^+ \qquad (\textit{with the convention } 0 = \frac{1}{\infty}).$$

$w(x,y)$ is the *weight* of the distance between the objects x and y, w is called a *weight function*.

In analogy to Degens (1983a) one can look for maxima of the likelihood function:

$$L(d_u \mid d) = \prod_{\substack{\{x,y\} \subseteq S \\ \sigma(x,y) < \infty}} \frac{1}{\sqrt{2\pi}\,\sigma(x,y)} \exp\left\{ -\frac{1}{2\sigma^2(x,y)}[d(x,y)-d_u(x,y)]^2 \right\} = Max!$$

For fixed $\sigma(x,y)$ this is equivalent to minimize the negative log likelihood function:

$$-\log L(d_u \mid d) \propto \sum_{\{x,y\} \subset S} \left[\frac{d(x,y)-d_u(x,y)}{\sigma(x,y)} \right]^2 = Min!$$

Furthermore this is equivalent to minimize the (weighted) residual sum of squares, cp. Degens (1985,1988):

$$\sum_{\{x,y\} \subset S} w(x,y)\left[d(x,y)-d_u(x,y)\right]^2 =$$

$$\sum_{C \in H(d_u)} \sum_{A,B \in N(C,H_0(d_u))} \sum_{x \in A, y \in B} w(x,y)\left[d(x,y)-d_u(x,y)\right]^2. \tag{4.1}$$

We denote the square root of this sum of squares for short by

$$\| d - d_u \|_w .$$

Hence the ML-principle leads as in Degens (1983,1985) to a special approximation problem:

$$\| d - d_u \|_w = Min ! \quad \{ d_u \text{ ultrametric } \}.$$

The same approximation problem arises (analogue to Degens 1983), if $\sigma(x,y)$ is known except of a constant factor, i.e. $\sigma(x,y) = c \sigma^*(x,y)$, with $\sigma^*(x,y)$ fixed, known, c unknown. The variances may be known from previous work, or they can be estimated. Degens & Lausen (1986) discussed the estimation in linear models. If the knowledge about the true hierarchy is poor, the approach of section 7 is preferable.

Note the the ML principle does not only justify the interest in global maxima, but also the interest in local maxima of the likelihood function. As shown above the approximation problem and the problem of maximizing the likelihood function are equivalent.

5. Approximation by Hierarchical Clustering

In the last section we state our interest to find a global proximum in the set $U(S)$ of all ultrametrics on S, i.e. for a given distance d we want to find an ultrametric d_u with

$$\| d_u - d \|_w \le \| d_u' - d \|_w \quad \forall d_u' \in U(S) .$$

Hartigan (1967) stated this weighted least squares approximation problem. Its computational difficulty is well known, and Krivanek (1986) proved its NP-completeness. There exist some proposals for approximate solutions by numerical methods (Carroll & Pruzansky, 1980; de Soete, 1984), but we want to consider a theoretically more substantiated approach: The computation of local proxima by hierarchical clustering methods.

As it is cumbersome to compute a global proximum, our interests are devoted to local proxima according to the following definition:

$d_u \in U(S)$ is a local proximum to a distance d with respect to $\| \ \|_w$, if there exists a neighbourhood U of d_u in $U(S)$ with

$$\| d_u - d \|_w \le \| d_u' - d \|_w \quad \forall d_u' \in U .$$

For the characterization of local proxima we introduce the following abbreviations:

Let w be a weight function.

For two disjoint and nonempty subsets A,B of S we define

$$w(A,B) := \sum_{a \in A, b \in B} w(a,b) .$$

By $D(S)$ we denote the set of all nonnegative distances on S. For $d \in D(S)$ we define the weighted average distance of all measured distances between A and B by

$$\bar{d}_w(A,B) := \frac{\sum\limits_{a \in A, b \in B} w(a,b)d(a,b)}{w(A,B)} \qquad \text{with } \frac{0}{0} := \infty \text{ for convenience.}$$

A simple characterization of local proxima is now given by the following theorem:

Characterization Theorem

Let $d \in D(S)$, and let w be a weight function.

Then $d_u \in U(S)$ is a local proximum to d with respect to $\| \ \|_w$, iff for all pairs A,B of lower neighbours in $H_0(d_u)$ of any cluster $C \in H_0(d_u)$ with $w(A,B) > 0$:

$$d_u^*(C) = \bar{d}_w(A,B) .$$

This theorem was first proved for unweighted distances by Degens (1985). The proof for weighted distances and some interesting extensions are given by Degens (1988). The sufficiency of the characterizing condition can be easily derived from the decomposition of $\| d - d_u \|_w$ given in (4.1).

Note that an ultrametric d_u is a local proximum to d, iff d_u is the global proximum in the set of all ultrametrics d_u' with $H(d_u')$ equal to or finer than $H(d_u)$.

It can be easily seen, that the well established agglomerative average linkage method (UPGMA)[†] yields a local proximum. Furthermore it is possible to generalize the average linkage method to a wide class of agglomerative hierarchical clustering methods, all resulting in local proxima. The main idea of this generalization can be described in the following way:

The usual average linkage method is an agglomerative method: In each step of the agglomerative process we have to choose two elements of the partition built in the previous step. These elements are joined to a new cluster. The usual average linkage method chooses in each step a pair A,B with minimal (weighted) average distance $\bar{d}_w(A,B)$.

---

†) The average linkage method is usually only defined for unweighted distances. But it can be easily adapted to weighted distances by substituting weighted averages for unweighted averages, cp. Degens (1988).

As this average distance also defines the level $d_u^*(A \cup B)$ of the new created cluster $A \cup B$, d_u^* is an isotonic function on the constructed hierarchy H and hence it corresponds to an ultrametric.

But it is possible to consider in each step several *admissible* pairs, characterized by a simple condition, so that the isotony is not violated, if in each step two elements of an admissible pair are joined. Hence selecting one of these admissible pairs in each step results in an ultrametric, and moreover this ultrametric is a local proximum, too.

The selection of an admissible pair can be done by minimizing a selection statistic in each step. Such a selection statistic must be defined on all admissible pairs in each step. It should depend on the distance d (and on the weight function w) and perhaps on additional information. We consider here only the additional dependence on the partition built in the previous step, which leads to the following definition of generalized agglomerative average linkage methods:

Definition

Let $\mathbb{P}(S)$ be the set of all partitions of S, and for $C \in \mathbb{P}(S)$ let $\mathbb{P}_2(C)$ be the set of all pairs of different elements of C.

Let $\Gamma := (g_C)_{C \in \mathbb{P}(S)}$ be a family of statistics g_C with

$$g_C \colon \mathbb{P}_2(C) \to \mathbb{R}_0^+ \quad \text{(additionally depending on } d \text{ and } w) \ .$$

Then the hierarchical clustering method φ_{AL}^Γ is defined by the following algorithm, which assigns a hierarchy H and a function $d_u^* \colon H \to \mathbb{R}_0^+$ to a distance $d \in D(S)$ and a weight function w :

begin
   $C_0 := \{\{x\} \mid x \in S\}$
   $H := C_0$
   $d_u^*(\{x\}) := 0 \ \forall x \in S$
   $i := 0$
repeat
   $i := i + 1$
   $A_i := \{\{A,B\} \in \mathbb{P}_2(C_{i-1}) \mid w(A,B) > 0 \ \wedge \ \forall C \in C_{i-1} - \{A,B\}$ holds:
      $\min(\bar{d}_w(A,C), \bar{d}_w(B,C)) < \bar{d}_w(A,B) \ \Rightarrow \ \bar{d}_w(A,B) < \bar{d}_w(A \cup B, C) \}$
    is the set of the admissible pairs
   Choose A,B minimizing $g_{C_{i-1}}$ among all pairs in A_i
   $C_i := (C_{i-1} - \{A,B\}) \cup \{A \cup B\}$
   $H := H \cup \{A \cup B\}$
   $d_u^*(A \cup B) := \bar{d}_w(A,B)$ defines the level of $A \cup B$
until $C_i = \{\{S\}\}$.

Remarks:

1) If we assume, that $w(A,\bar{A})>0$ for all bipartitions $\{A,\bar{A}\}$ of S, then there exists at least one admissible pair of sets in each step, because each pair $\{A,B\}$ minimizing $\bar{d}_w$ on $\{\{A,B\}\in\mathbb{P}_2(C_{i-1})\mid w(A,B)>0\}$ is admissible and this set is not empty.

2) φ^{Γ}_{AL} may be not unique, if several admissible pairs minimize $g_{C_{i-1}}$ in the i-th step. We neglect the problem how to decide in such situations, but the following lemmata about methods φ^{Γ}_{AL} hold for any arbitrary decision.

3) If $g_C(A,B)=\bar{d}_w(A,B)$ then φ^{Γ}_{AL} corresponds to the usual agglomerative average linkage method.

We first show, that the function d^*_u on H, both constructed by φ^{Γ}_{AL}, is isotonic.

Lemma

Let Γ be according to the above definition.
Let be $d\in D(S)$ and let w be a weight function.
Let H and d^*_u be constructed by φ^{Γ}_{AL} applied to d and w. Then

$$d^*_u(A)\le d^*_u(B) \quad \forall A,B\in H \text{ with } A\subseteq B$$

holds.

Proof:

Let $A,B\in H$ with $A\subseteq B$.
It suffices to show that $d^*_u(A)\le d^*_u(B)$ if A is a lower neighbour of B.
If A is minimal in H we have $d^*_u(A)=0$ and hence $d^*_u(A)\le d^*_u(B)$, because d is non-negative.
Otherwise let $C_{i-1}=\{C_1,...,C_p\}$ be the partition before the i-th step, where A is created, and let $C_{j-1}=\{D_1,...,D_q\}$ be the partition before the j-th step, where B is created. We can assume without loss of generality $A=C_1\cup C_2$, $B=D_1\cup D_2$, $A=D_1$ and, as $i<j$, $D_2=\bigcup_{k=3}^{l} C_k$ with $3\le l\le p$. As $\{C_1,C_2\}$ is admissible in the i-th step, we have

$$d^*_u(A)=\bar{d}_w(C_1,C_2)\le\bar{d}_w(C_1\cup C_2,C_k)=\bar{d}_w(D_1,C_k) \quad \forall 3\le k\le p.$$

Furthermore we have

$$d^*_u(B)=\bar{d}_w(D_1,D_2)\ge\min\{\bar{d}_w(D_1,C_k)\mid 3\le k\le l\}.$$

Hence we have $d^*_u(A)\le d^*_u(B)$.

Therefore we can regard each method φ^{Γ}_{AL} as a mapping, assigning the ultrametric $\varphi^{\Gamma}_{AL}(d,w)$ to each pair of a distance d and a weight function w.

Next we show, that each φ^{Γ}_{AL} results in a local proximum.

Lemma

Let Γ be according to the above definition.

Let be $d \in D(S)$ and let w be a weight function.

Then the ultrametric $\varphi^{\Gamma}_{AL}(d,w)$ is a local proximum to d resp. $\| \|_w$.

Proof:

Let H be constructed by φ^{Γ}_{AL} applied to d and w, and let $H' := H_0(d_u)$ with $d_u := \varphi^{\Gamma}_{AL}(d,w)$. Then $H' \subseteq H$ and for $A \in H'$ the level $d^*_u(A)$ corresponds to the value assigned by φ^{Γ}_{AL}.

Let A be a nonminimal member of H', i.e. $N(A,H') \neq \phi$.

Let A be created in the k-th step ($k \geq 1$). Then each $C_i' := \{ C \in C_i \mid C \subseteq A \}$ for $0 \leq i < k$ is a partition of A. For each $C \in C_i'$ there exists $N \in N(A,H')$ with $C \subseteq N$, or there exist $N_1,...,N_l \in N(A,H')$ with $C = \bigcup\limits_{j=1}^{l} N_j$. Hence the set D_i of all maximal elements in $N(A,H') \cup C_i$ is a partition of A, too.

Now we will show

$$\forall 0 \leq i < k \ \forall D_1, D_2 \in D_i \text{ with } w(D_1,D_2) > 0 \text{ holds } \bar{d}_w(D_1,D_2) = d^*_u(A) \ . \qquad (5.1)$$

We prove (5.1) per induction over i, starting with $i = k - 1$.

$i = k - 1$:

$D_{k-1} = \{ D_1, D_2 \}$ is the set of the lower neighbours of A in H. D_1 and D_2 are joined in the k-th step, resulting in the set A. Hence we have $\bar{d}_w(D_1,D_2) = d^*_u(A)$.

$i \rightarrow i-1$:

If $D_i = D_{i-1}$ there is nothing to prove.

Otherwise there exist $C,C' \in D_{i-1}$ with $D_i = (D_{i-1} - \{ C,C' \}) \cup \{ C \cup C' \}$, and C and C' are joined in the i-th step. The latter implies $d^*_u(C \cup C') = \bar{d}_w(C,C')$. But $d^*_u(C \cup C') = d^*_u(A)$ holds, too, because $C \cup C'$ implies $d^*_u(C \cup C') \leq d^*_u(A)$ and $d^*_u(C \cup C') < d^*_u(A)$ implies the existence of $N \in N(A,H')$ with $C \cup C' \subseteq N$, which contradicts $C,C' \in D_{i-1}$. So we have $d^*_u(A) = \bar{d}_w(C,C')$ and it remains to show

$$\bar{d}_w(C,F) = d^*_u(A) \ \ \forall F \in D_{i-1} - \{ C,C' \} \text{ with } w(C,F) > 0 \ .$$

Let $F \in D_{i-1} - \{C, C'\}$ with $w(C,F) > 0$. There exist $E_1, ..., E_l \in C_{i-1}$ with $F = \bigcup_{j=1}^{l} E_j$.

For $1 \le j \le l$ with $w(C,E_j) > 0$ the admissibility of $\{C,C'\}$ implies

$$\bar{d}_w(C \cup C', E_j) \ge \bar{d}_w(C,C')$$
$$\wedge \quad \bar{d}_w(C \cup C', E_j) = \bar{d}_w(C,C') \Rightarrow \bar{d}_w(C,E_j) = \bar{d}_w(C,C') .$$

As $C \cup C'$ and F are members of D_i, by induction we have

$$d_u^*(A) = \bar{d}_w(C \cup C', F) = \frac{1}{w(C \cup C', F)} \sum_{\substack{1 \le j \le l \\ w(C \cup C', E_j) > 0}} w(C \cup C', E_j) \bar{d}_w(C \cup C', E_j).$$

Hence $\bar{d}_w(C \cup C', E_j) > \bar{d}_w(C,C') = d_u^*(A)$ for $j \in \{1, ..., l\}$ with $w(C \cup C', E_j) > 0$ is impossible. Consequently we have $\bar{d}_w(C, E_j) = \bar{d}_w(C,C') = d_u^*(A) \forall 1 \le j \le l$ with $w(C, E_j) > 0$, which implies $\bar{d}_w(C,F) = d_u^*(A)$.

Now (5.1) implies the condition of the characterization theorem for $i=0$, because $D_0 = N(A, H_0(d_u))$.

So far we present a wide class of generalized average linkage methods φ_{AL}^{Γ} to compute local proxima. The usual average linkage method is the most established member of this group. But an appropriate choice of Γ may result in a better method.

We now present a proposal for Γ, which is based on considerations first mentioned in Vach & Degens (1988a).

Let us assume, that the partition C, built in a step of the agglomeration process, includes only clusters of the true hierarchy, i.e.

$$C \in H(d_u) \forall C \in C$$

holds.

Then for $A, B \in C$ the ultrametric distance $d_u(a,b)$ is constant for $a \in A, b \in B$. If $d_u(A,B)$ denotes this constant, then all these values define an ultrametric on C, i.e. we have

$$d_u(A,B) \le \max\{d_u(A,C), d_u(B,C)\} \quad \forall A,B,C \in C .$$

Now the union of two elements A, B of C results in a cluster of $H(d_u)^{\dagger}$ iff

$$d_u(A,B) \le d_u(A,C) = d_u(B,C) \quad \forall C \in C - \{A,B\} . \tag{5.2}$$

---

†) If $H(d_u)$ is not maximal fine, the new cluster $A \cup B$ may be only *compatible* with $H(d_u)$, i.e $\{A \cup B\} \cup H(d_u)$ is a hierarchy.

As d_u is unknown it is straightforward to replace d_u by $\bar{d}_w$ in (5.2). The usual average linkage method minimizes the average distance $\bar{d}_w(A,B)$ in order to satisfy the inequality in (5.2), but it neglects the information in the data according to the equality in (5.2).

For generalized average linkage methods the admissibility of a pair A,B can be regarded as a check on the inequality in (5.2). Hence an appropriate selection statistic should mainly check the equality in (5.2). It is straightforward to base such a check on the differences

$$\bar{d}_w(A,C) - \bar{d}_w(B,C) \quad \text{with } C \in C - \{A,B\} \ .$$

We propose to check the equality in (5.2) by minimizing the average of the squares of the standardized differences, i.e. we propose to use $\varphi_{AL}^{\Gamma*}$ with

$$g_C^*(A,B) = \begin{cases} \infty & \text{if } C_{A,B}^{>} = \phi \\ \dfrac{\sum\limits_{C \in C_{A,B}^{>}} (\Lambda(A,B,C))^2}{|C_{A,B}^{>}|} & \text{if } C_{A,B}^{>} \neq \phi \end{cases}$$

$$\text{with} \quad \Lambda(A,B,C) := \frac{\bar{d}_w(A,C) - \bar{d}_w(B,C)}{\sqrt{1/w(A,C) + 1/w(B,C)}}$$

$$\text{and} \quad C_{A,B}^{>} := \{ C \in C - \{A,B\} \mid w(A,C) > 0 \wedge w(B,C) > 0 \} \ .$$

An alternative derivation for g_C^* based on an maximum likelihood approach is given by Lausen (1988). If all weights are greater than 0, then $|C_{A,B}^{>}|$ is constant and $|C_{A,B}^{>}| g_C^*(A,B)$ can be regarded as the increase of the weighted residual sum of squares caused by the union of A and B (see Vach & Degens, 1988b,c).

We denote the corresponding generalized average linkage method $\varphi_{AL}^{\Gamma*}$ for short by φ_{AL}^{new}. The example in the next section demonstrates, that φ_{AL}^{new} improves the usual average linkage method φ_{AL}^{old} in a remarkable degree, at least for distances on larger sets. An extensive comparison of both methods is given in Vach & Degens (1988b).

Considering the maximum likelihood approach in the last section local proxima correspond to local maxima of the likelihood function. Of course there is no guarantee that a local maximum of the likelihood function is a global one, in general. But asymptotically - for $e \to 0$ - there is only one local maximum for dichotomic ultrametrics d_u. Regarding the assumptions in the last section, both φ_{AL}^{old} and φ_{AL}^{new} (and all generalized average linkage methods, too) are asymptotically efficient estimators, as known from the statistical theory of linear models.

6. Judgement and Comparison of Hierarchical Clustering Methods

Using hierarchical clustering methods to estimate in a stochastic framework, it is possible to apply statistical methods to judge the goodness of a hierarchical clustering method. Considering the general error model (3.1) a measure for the goodness of a hierarchical clustering method $\varphi: D(S) \to U(S)$ depends on the distribution of $\varphi(d)$, i.e. on the true ultrametric d_u and the error distribution F (a weight function w may be included in F). As the goodness of a method may rather vary for different ultrametrics, it is necessary to consider several fixed ultrametrics in order to examine the variation. A random generation of ultrametrics (sometimes proposed in the literature) seems to be inappropriate for this use.

A sensible measure for the goodness of a hierarchical clustering method φ may measure the expected deviance between $\varphi(d)$ and the true ultrametric d_u. But measures like the expected squared Euclidian distance

$$E_{F,d_u} \| \varphi(d) - d_u \|_2^2$$

are perhaps to smooth, because partial goodness of a hierarchical clustering method like good redetection of small clusters may be covered by such a measure. Because we are mainly interested in reconstructing the true phylogeny, i.e. the true hierarchy $H(d_u)$, an alternative and more appropriate measure is given by the probability to (re-)detect a true cluster $C \in H(d_u)$ with φ:

$$DP(\varphi, C, d_u, F) := P_{F, d_u}(C \in H(\varphi(d))) \ .$$

This measure was originally proposed by Ostermann & Degens (1984,1986). The different redetection probabilities for different clusters C allow to judge the partial goodness of a hierarchical clustering method. Moreover redetection probabilities are easy to interpret and allow to judge a single method, whereas the expected squared Euclidian distance allows only to compare several methods.

In general it seems to be impossible to evaluate redection probabilities by analytic methods. But it is easy to approximate these probabilities with a Monte Carlo study: We generate independent realizations of d according to the considered error model and we observe the frequency to find C in the hierarchies constructed by the method φ applied on d. Then the redetection frequency approximates the redetection probability. Analogously the average squared Euclidian distance between the ultrametric $\varphi(d)$ and d_u approximates the expected squared Euclidian distance.

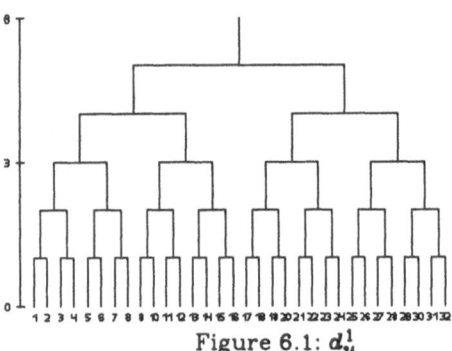

Figure 6.1: d_u^1

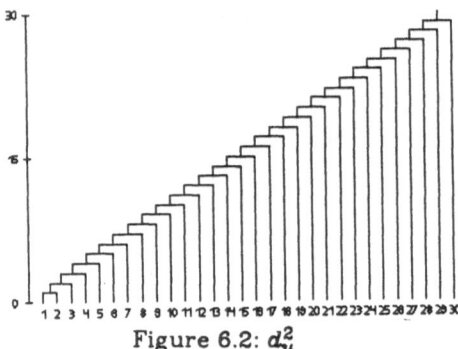

Figure 6.2: d_u^2

To illustrate the use of the two proposed measures we compare the methods φ_{AL}^{old} and φ_{AL}^{new}, proposed in the previous section. We consider two ultrametrics d_u^1 and d_u^2, which are given by the dendrograms in figure 6.1 and 6.2, and we consider three families of error distributions, which are all based on the assumption of independent errors:

$$F_\sigma^1: e(x,y) \sim N(0,\sigma^2)$$
$$F_\sigma^2: e(x,y) \sim N(0,(d_u(x,y)\sigma)^2)$$
$$F_\sigma^3: e(x,y) \sim 0.95\,N(0,\sigma^2) + 0.05N(0,(3\sigma)^2)$$

The heteroscedastic model F_σ^2 is sometimes considered for the reconstruction of evolutionary trees. The gross-error-model F_σ^3 allows to consider the outlier sensitivity of the methods (compare Vach & Degens, 1986). For each family of error distributions we examine several values of the scale parameter.

Our Monte Carlo study is based on 1000 independent pseudo-random generations of the random variable d for each pair of considered ultrametric and error distribution. We first look at the results for one single scale parameter for each pair of ultrametric and family of error distribution; this scale parameter is chosen to obtain a redection frequency with φ_{AL}^{old} of about 50% for the largest clusters. In order to compare φ_{AL}^{old} and φ_{AL}^{new} table 6.1 presents the average squared Euclidian distances and table 6.2 presents the average redetection frequencies of the clusters of size 16, 8, 4 and 2 in $H(d_u^1)$ (clusters of equal size possess equal redetection probabilities in this case) and the redetection frequencies of the clusters of size 29, 20, 11 and 2 in $H(d_u^2)$. Comparing the average squared Euclidian distances we find remarkable improvements with φ_{AL}^{new} for all considered cases. But table 6.2 gives a more detailed impression: Again we observe remarkable improvements in redetecting large clusters, but (low) deteriorations appear for small clusters. Obviously such deteriorations for small clusters cannot affect the expected squared Euclidian distance. Nevertheless the results demonstrate the superiority of φ_{AL}^{new} in the considered cases.

Comparing the results between the six considered cases, the improvement in redection probability is more distinct for d_u^1 than for d_u^2. Restricted to d_u^1 the improvement is more distinct for F^2 and F^3 than for F^1. The latter demonstrates that φ_{AL}^{new} is not more based on the assumption of normal errors than φ_{AL}^{old}.

To widen our examination to other scale parameters we give graphical representations of the redetection frequencies of the largest clusters for a range of scale parameters in figure 6.3. For the six considered cases we plot the redetection frequency of the largest clusters versus the scale parameter and connect the plotted points by lines: a solid line for the redetection frequencies of φ_{AL}^{new} and a dotted line for the redetection frequencies of φ_{AL}^{old}. To improve the comparability of the plots we standardize the scale parameter σ so that $\sigma=1$ corresponds to a redetection frequency of 50% by φ_{AL}^{old}.

The plots demonstrate again the remarkable degree of improvement by φ_{AL}^{new}. Note that the smaller increase of redetection probability for d_u^2 than for d_u^1 may be due to the slower decreasing of the corresponding curve.

		d_u^1			d_u^2	
	σ	φ_{AL}^{old}	φ_{AL}^{new}	σ	φ_{AL}^{old}	φ_{AL}^{new}
$e \sim F_\sigma^1$	1.6	285.8	222.9	3.1	1163.8	825.7
$e \sim F_\sigma^2$	0.68	184.0	105.1	0.57	687.8	423.5
$e \sim F_\sigma^3$	0.97	200.5	116.5	2.5	1363.7	906.8

Table 6.1: average squared Euclidian distances

	number of objects	$e \sim F_{1.6}^1$		$e \sim F_{0.68}^2$		$e \sim F_{0.97}^3$	
		φ_{AL}^{old}	φ_{AL}^{new}	φ_{AL}^{old}	φ_{AL}^{new}	φ_{AL}^{old}	φ_{AL}^{new}
d_u^1	16	49.9	86.6	49.7	96.8	49.5	95.7
	8	26.5	34.0	38.6	61.2	44.7	69.8
	4	26.8	14.8	32.7	31.5	37.0	37.8
	2	31.9	28.9	43.8	40.3	43.7	40.8
	number of objects	$e \sim F_{3.1}^1$		$e \sim F_{0.57}^2$		$e \sim F_{2.5}^3$	
		φ_{AL}^{old}	φ_{AL}^{new}	φ_{AL}^{old}	φ_{AL}^{new}	φ_{AL}^{old}	φ_{AL}^{new}
d_u^2	29	50.1	61.6	49.8	66.3	50.3	60.7
	20	30.3	42.9	37.7	50.4	30.9	39.2
	11	26.7	28.0	48.8	43.7	27.2	30.0
	2	28.2	24.4	68.6	51.7	28.6	25.7

Table 6.2: frequencies of redetection (in %)

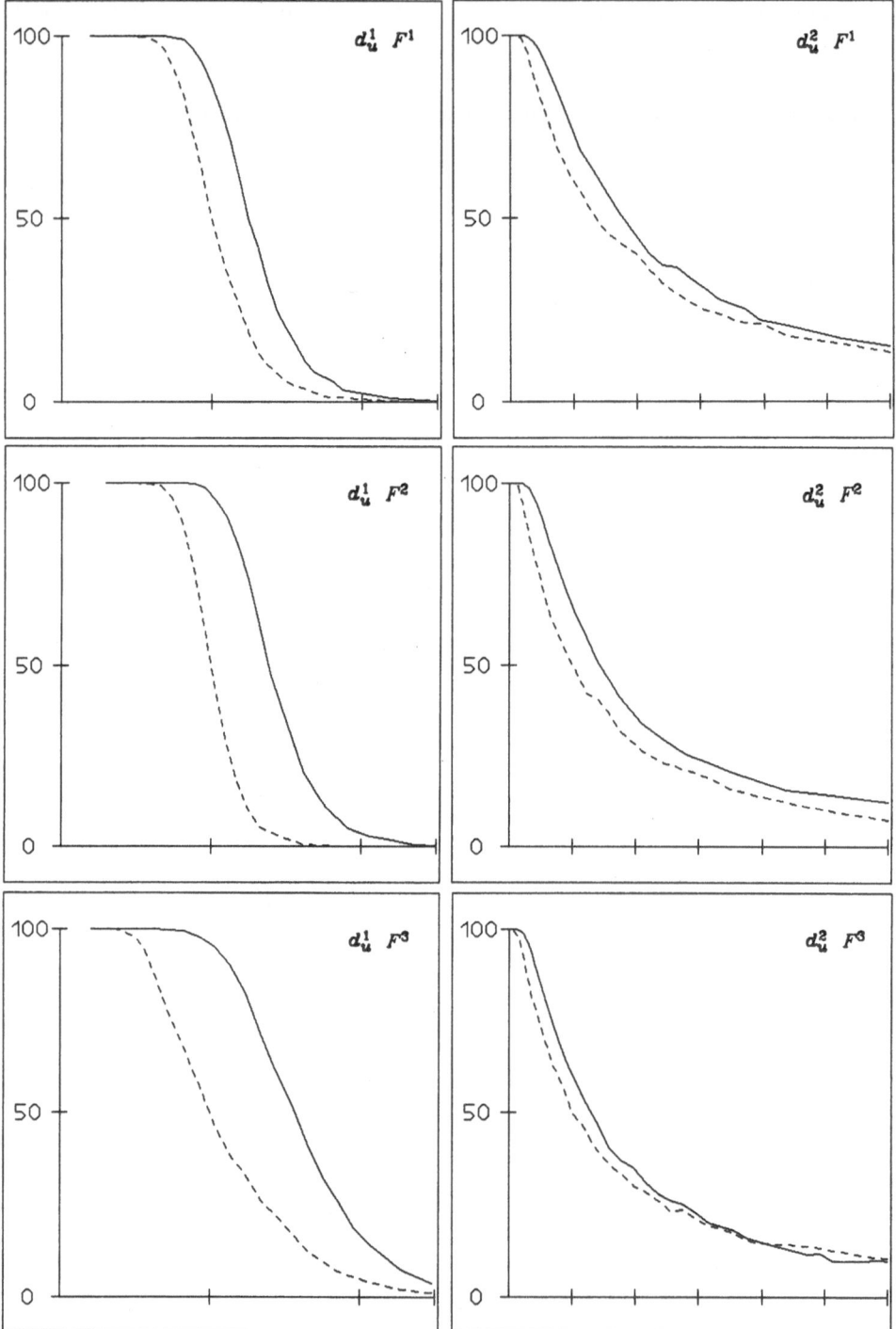

Figure 6.3: plots of redetection frequencies

7. Accuracy and Stability of Estimated Phylogenies

Having obtained, i.e. estimated, a phylogeny the problem arises to evaluate the accuracy and stability of the phylogeny and its induced hierarchy. Therefore Lausen & Degens (1986) suggested variance estimation based on subsets consisting of three objects of S. Variance estimation by the residual sum of squares depends strongly on the estimated hierarchy and will differ from one local proximum to another, but evaluation methods in hierarchical cluster analysis based on three objects methods do not suffer from this dependency. In section 5 we define reconstruction methods for a distance on S, consequently we use as measured dissimilarities the means of the replicates. The variation of the means of the replicates provides an important information for the evaluation of a provisional reconstruction. For the variation characteristic, the variance of the means $\bar{d}(x,y)$, model (3.3) implies

$$Var(\bar{d}(x,y)) = \sigma_0^2 + \frac{\sigma_1^2}{r(x,y)} = \frac{\sigma^2}{w(x,y)} \quad \text{with } w(x,y), \sigma^2 > 0 . \qquad (7.1)$$

For the following $w(x,y)$ is an appropriate weight of the mean $\bar{d}(x,y)$ (cp. (3.4) and (4.1)). Note, that it is possible to define appropriate weights as a function of the number of replicates and the ratio of the variance components. Let $\bar{d}_{(i)}$; $i=1,2,3$ denote the order statistics of the three independently distributed means, $\bar{d}(x',y')$, $\bar{d}(x',z')$, $\bar{d}(y',z')$, between three objects $\{x',y',z'\} \subset S$, and let w_i denote the corresponding weights. Without loss of generality we denote x',y',z' by x,y,z with $d_u(x,y) \leq d_u(x,z) = d_u(y,z)$ and we define $\Delta(x,y,z) = d_u(x,z) - d_u(x,y)$. The agglomerative average linkage method φ_{AL}^{old} gives following three objects estimators:

$$\hat{d}_u(x,y) = \bar{d}_{(1)} , \quad \hat{d}_u(x,z) = \hat{d}_u(y,z) = \frac{w_2 \bar{d}_{(2)} + w_3 \bar{d}_{(3)}}{w_2 + w_3} \quad \text{and} \qquad (7.2)$$

$$\hat{\Delta}(x,y,z) = \hat{d}_u(x,z) - \hat{d}_u(x,y) .$$

A straightforward three objects variance estimator is given by:

$$\hat{\sigma}_{3ob}^2(\{x,y,z\}) = \frac{w_2 w_3}{w_2 + w_3}(\bar{d}_{(3)} - \bar{d}_{(2)})^2 = \sum_{i=2}^{3} w_i(\bar{d}_{(i)} - \hat{d}_u(x,z))^2 . \qquad (7.3)$$

Methods based on the three objects estimators (7.2) and (7.3) are called *three objects methods* (Lausen & Degens, 1988). In the case of *missing values* we define three objects methods on the subsets $\{x,y,z\}$ with three measured means; i.e. subsets without missing values. For notational convenience we present our formulae neglecting missing values in the present section, but the changes are obvious in case of missing values. In practice the weights and the variance are unknown (see (7.1)). Consequently we use (7.3) for a variance estimator in the balanced case of model (3.3) and for a variance component estimator in the unbalanced case of model (3.3) (cp. Lausen & Degens, 1986 or Lausen, 1987).

In the *balanced case* an interesting statistical characteristic of the variance estimator (7.3) is its expectation V as a function of Δ / σ :

$$V(\Delta / \sigma) = E(\hat{\sigma}^2_{3Ob}(\{x,y,z\})) \qquad \text{for } \Delta \geq 0 . \tag{7.4a}$$

Lausen & Degens (1986) gave an integral representation of $V(\Delta / \sigma)$. For convenience we use an approximation of $V(\Delta / \sigma)$ based on a Monte Carlo study with 100 000 repetitions for each $\Delta / \sigma \in \{0.0, 0.01, 0.02,..., 1.99, 2.0, 2.1,..., 19.9\}$. We derived by regression the approximation:

$$V(\Delta / \sigma) \approx 1 - 0.1943096 \, \tau(\Delta / \sigma - 1) \tag{7.4b}$$

$$+ 0.1569990 \, \tau(\Delta / \sigma - 0.5)$$
$$- 0.1908670 \, \tau(\Delta / \sigma)$$
$$- 0.5101694 \, \tau(\Delta / \sigma + 0.5)$$
$$+ 0.4002337 \, \tau(\Delta / \sigma + 1)$$
$$- 0.2485101 \, \tau(\Delta / \sigma + 2)$$

where $\tau(\Delta) = exp(-0.5 \, \Delta^2)$. The maximum absolute difference between (7.4b) and our Monte Carlo computation was about 0.011.

A variance estimator $\hat{\sigma}^2_{3Ob}(S)$ using all different subsets of three objects is the arithmetic mean of the subset estimators:

$$\hat{\sigma}^2_{3Ob}(S) = \frac{1}{\binom{n}{3}} \sum_{\{x,y,z\} \subset S} \hat{\sigma}^2_{3Ob}(\{x,y,z\}) . \tag{7.5}$$

It is possible to reduce the bias of $\hat{\sigma}^2_{3Ob}(S)$ by a weighted version. We include here the definition of a *weighted three objects variance estimator* of Lausen (1987) based on proposals of Lausen & Degens (1986). The weighted three objects variance estimator $\hat{\sigma}^2_{W3Ob}$ is defined as limit of $\hat{\sigma}^{2^i}_{W3Ob}$ for $i \to \infty$ of the following iterative procedure:

$$\hat{\sigma}^{2^1}_{W3Ob}(S) = \infty \quad \text{and} \quad \hat{\sigma}^{2^i}_{W3Ob}(S) = f(\hat{\sigma}^{2^{i-1}}_{W3Ob}(S)) \text{ , where} \tag{7.6}$$

$$f(a) = \frac{\displaystyle\sum_{\{x,y,z\} \subset S} W\left[\hat{\Delta}(x,y,z), \hat{\sigma}^2_{3Ob}(S)\right] B\left[\hat{\Delta}(x,y,z), a\right] \hat{\sigma}^2_{3Ob}(\{x,y,z\})}{\displaystyle\sum_{\{x,y,z\} \subset S} W\left[\hat{\Delta}(x,y,z), \hat{\sigma}^2_{3Ob}(S)\right]} .$$

Due to Monte Carlo studies and some heuristics (cp. figure 7.1 and Lausen, 1987) we use the following bias correction B and weights W:

$$B\left[\hat{\Delta}(x,y,z), \hat{\sigma}^{2^{i-1}}_{W3Ob}(S)\right] = \tag{7.7}$$

$$\frac{1}{V\left[\left[1 - \exp\left[-0.4 \frac{(\hat{\Delta}(x,y,z))^2}{\hat{\sigma}^{2^{i-1}}_{W3Ob}(S)}\right]\right] \frac{\hat{\Delta}(x,y,z)}{\sqrt{\hat{\sigma}^{2^{i-1}}_{W3Ob}(S)}}\right]} .$$

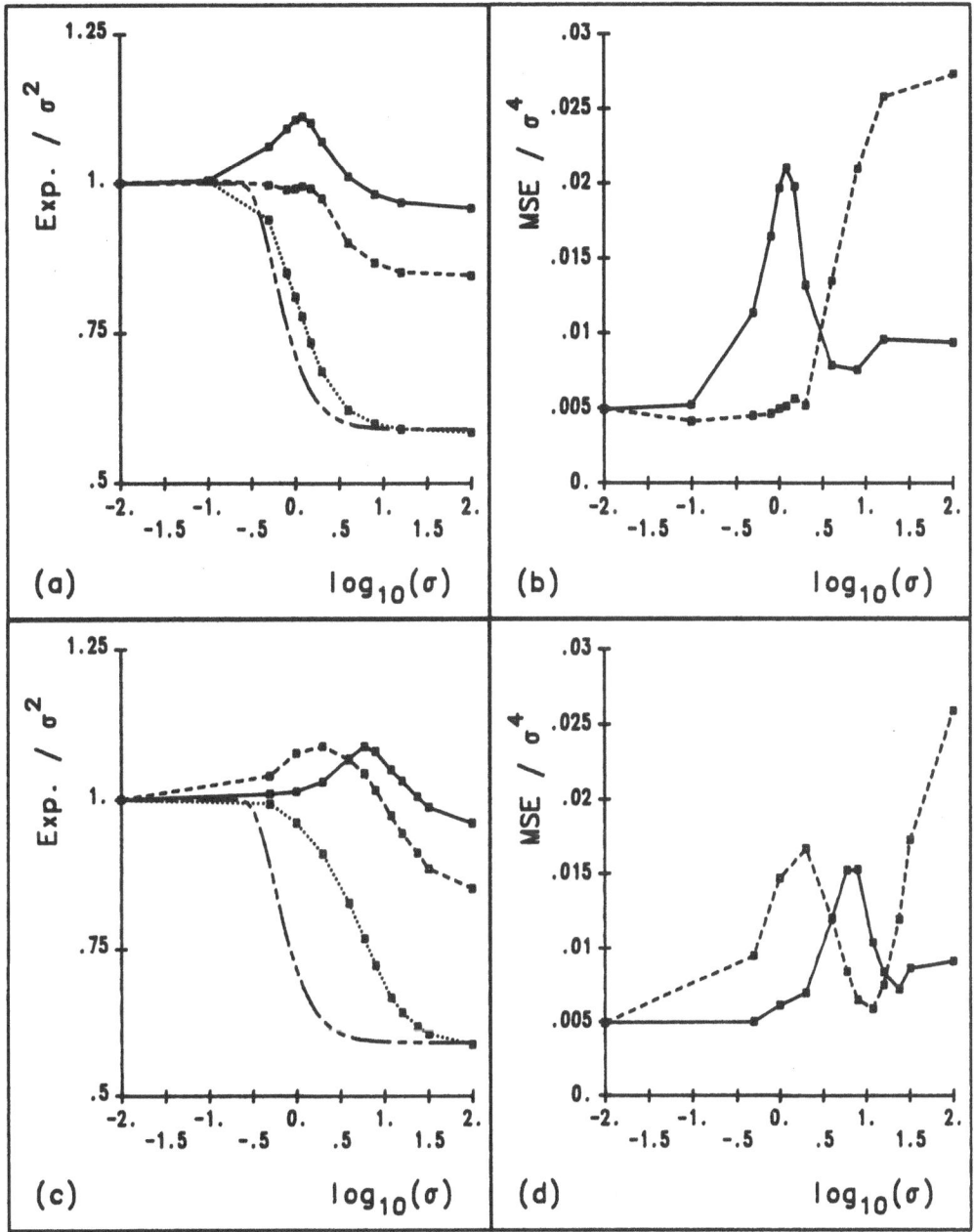

Figure 7.1: Monte Carlo results for variance estimators. The simulated expectation and MSE (symbol square) are connected by straight lines. $\hat{\sigma}^2_{3Ob}$: dotted; $\hat{\sigma}^2_{W3Ob}$: solid; variance estimation by φ^{old}_{AL} estimate of design matrix: dashed; approximation of $V(1\,/\,\sigma)$: long and short dashed. (a) and (b) are computed under d^1_u. (c) and (d) are computed under d^2_u for $n = 32$.

$$W\left[\hat{\Delta}(x,y,z),\ \hat{\sigma}^2_{0_{3Ob}}(S)\right] = \left[1 - \exp\left[-0.2\ \frac{\hat{\Delta}(x,y,z)}{\sqrt{\hat{\sigma}^2_{0_{3Ob}}(S)}}\right]\right].$$

V being the expectation as defined in (7.4) and reproduced in figure 7.1a,c. Figure 7.1 allows a comparison of Monte Carlo results for the different variance estimators. Moreover figure 7.1 covers the approximation (7.4b) and the classical variance estimator in linear models based on the φ^{old}_{AL} estimate of the design matrix. We simulated the expectation and mean square error (MSE) for two different ultrametrics with 1000 repetitions: The highly symmetric ultrametric d^1_u of section 6 (cp. figure 6.1) and the chain ultrametric d^2_u of section 6 for $n = 32$.

In the *unbalanced case* a nonnegative three objects estimator for $\sigma_0{}^2$ is given by:

$$\hat{\sigma}^2_{0_{3Ob}}(\{x,y,z\}) = \frac{t^2}{t^2 + 1}\ \frac{(\bar{d}_{(3)} - \bar{d}_{(2)})^2}{2}\ , \tag{7.8}$$

where r_i denotes the number of replicates of $\bar{d}_{(i)}$ and $t = 2\,r_2\,r_3\,/\,(r_2 + r_3)$. The choice of t gives in the balanced case for $\Delta \to \infty$ a nonnegative variance component estimator introduced by Hartung (1981). Therefore we receive the variance component estimator

$$\hat{\sigma}^2_{0_{3Ob}}(S) = \frac{1}{\binom{n}{3}}\ \sum_{\{x,y,z\}\subset S}\ \hat{\sigma}^2_{0_{3Ob}}(\{x,y,z\})\ . \tag{7.9}$$

Modifying the estimator of $\Delta(x,y,z)$ for the unbalanced case to

$$\hat{\Delta}'(x,y,z) = \frac{\hat{d}_u(x,z) - \hat{d}_u(x,y)}{\sqrt{\dfrac{2}{3}\left[\dfrac{1}{w_2 + w_3} + \dfrac{1}{w_1}\right]}}\ , \tag{7.10}$$

we provide a weighted estimator $\hat{\sigma}^2_{0_{W3Ob}}(S)$ by replacing $\hat{\Delta}$ with $\hat{\Delta}'$ and $\hat{\sigma}^2_{0_{3Ob}}(\{x,y,z\})$ with $\hat{\sigma}^2_{0_{3Ob}}(\{x,y,z\})$ in the definition of $\hat{\sigma}_{W3Ob}$. The classical estimator of $\sigma_1{}^2$ is:

$$\hat{\sigma}_1{}^2 = \frac{1}{\sum\limits_{\{x,y\}\subset S} r(x,y) - \binom{n}{2}}\ \sum_{\{x,y\}\subset S}\ \sum_{i=1}^{r(x,y)} (d(x,y,i) - \bar{d}(x,y))^2\ . \tag{7.11}$$

For the following we define the weight $w(x,y)$ (formula (7.1)) as reciprocal to the variance. We propose to use estimated weights given by the nonnegative weighted three objects variance estimator $\hat{\sigma}^2_{0_{W3Ob}}(S)$ for $\sigma_0{}^2$ and the classical unbiased estimator $\hat{\sigma}_1{}^2$ in the unbalanced case. Note the weights are redundant in the balanced case.

Suggesting the detection probability $DP(\varphi,C,d_u,F)$ (cp. section 6) as stability criterion Lausen & Vach (1986) used the bootstrap idea and got the estimator

$$\widehat{DP}(\varphi, C, d_u, F_{\sigma^2}) = DP(\varphi, C, \hat{d}_u, F_{\hat{\sigma}^2}) \ . \tag{7.12}$$

For example Efron (1982) or Rothe (1988) discussed bootstrap methods in general. Felsenstein (1985) suggested a bootstrap method for multivariate observations per species. Dubes & Jain (1979) gave a review of other approaches to cluster validity.

The proposal (7.12) is based on an estimator $\hat{d}_u$ of the ultrametric and a variance estimator $\hat{\sigma}^2$. Consequently we have the problem, that $\widehat{DP}$ may tend to overestimate DP in the case of estimated clusters; i.e. $C \in H(\hat{d}_u)$. Trying to reduce this problem, we use three objects variance estimation. The underlying idea of the bootstrap proposal is to use the obtained variability of the data to evaluate the estimated, i.e. obtained, hierarchy regarding the used hierarchical clustering method. Lausen (1988) argued that his nonisotonic clustering method tends to reduce the bias of $\hat{d}_u$ and therefore improves the bootstrap approach. In the *unbalanced case* the variance structure is determined by a diagonal matrix $diag(\sigma_0^2 + \sigma_1^2 / r(x,y))$ and we use for $\widehat{DP}$ the estimate $diag(\widehat{\sigma_0^2}_{W3Ob}(S) + \widehat{\sigma_1^2}/ r(x,y))$.

On the basis of Monte Carlo results for φ_{AL}^{old} (Lausen & Degens, 1987; Lausen, 1987) we propose to evaluate estimated clusters of size greater than two with the bootstrap idea as follows:

$\widehat{DP} > 0.95$ *high stability* of the estimated cluster C,

$0.95 \geq \widehat{DP} > 0.90$ *stability* of the estimated cluster C,

$0.90 \geq \widehat{DP} > 0.80$ *some stability* of the estimated cluster C,

$0.80 \geq \widehat{DP}$ *no stability* of the estimated cluster C.

We summarize, that variance and variance components estimation allows to evaluate the accuracy of the measured distance data and the computation of a bootstrap measure of fit for estimated clusters. For a fixed hypothetical cluster the approach gives a bootstrap estimator of the detection probability.

8. Model Verification

The verification of the structural and stochastical assumptions is an important part of the statistical analysis. The structural setting is discussed in section 2, therefore we emphasize once more, that the approximation of the phylogeny by a mathematical graph, i.e. a tree, has to be biologically reasonable. The more general structure is the additive tree model. To check the necessity of such a model, Lausen (1987) and Lausen & Degens (1988) suggested following procedure: For model (3.2) we receive for $\Delta(x,y,z) \to \infty$

$$E(\widehat{\sigma}^2_{3Ob}(\{x,y,z\})) = \sigma^2 + \frac{w_2 w_3}{w_2 + w_3} 2 \sigma_\gamma^2 d_u(x,y) \ . \tag{8.1}$$

Due to the asymptotic argument of formula (8.1) and Monte Carlo results (cp. Lausen, 1987), Bravais-Pearsons empirical correlation coefficient $\hat{\rho}$ between $(\hat{\sigma}^2_{30b}(\{x,y,z\}))_{\{x,y,z\} \subset S}$ and $(w_2 w_3 / (w_2 + w_3) \, \hat{d}_u(x,y))_{\{x,y,z\} \subset S}$ provides a three objects method to evaluate the influence of the covariance component σ_γ^2. A positive correlation indicates a possible influence of the fluctuations of the evolution, i.e. of σ_γ^2.

A *lack of fit test* provides another check for the necessity of an additive tree model. Consequently we suggest to test the fit of the ultrametric vs. the additive tree under the estimated hierarchy induced by the used clustering procedure. But we emphasize that such a lack of fit test and the proposed residual analysis below depends on the estimated hierarchy and therefore we work under a conditional linear model.

Using the estimated design, we propose standard techniques of *residual analysis* in linear models (e.g. Cook & Weisberg, 1982) to check the stochastical models introduced in section 3. In this context residuals are the differences of the observed values and the fitted values; e.g. $\bar{d} - \hat{d}_u$. Therefore graphical checking involves a plot of the standardized residuals vs. the fitted values and a quantile quantile plot (Q-Q-plot) of the standardized residuals. Moreover the pattern of the residuals may be a hint for a deviation of the estimated hierarchy from the true hierarchy.

9. Example: New World Suboscine Passerine Birds

Sibley & Ahlquist (1985) analysed the phylogeny of *New World Suboscine Passerine Birds* with DNA-DNA hybridization data. They measured 967 $\Delta T_{50}H$ dissimilarities between more than 100 species of birds. Sibley & Ahlquist (1983) discussed the method $\Delta T_{50}H$. Sibley & Ahlquist (1985) used 16 different species as *tracer*-species in the hybridization experiments. The tracer-species are (numbers in brackets are used instead of the names later on):

Myiarchus tyrannulus (Brown-crested Flycatcher) [1],
Sayornis phoebe (Eastern Phoebe) [2],
Elaenia frantzii (Mountain Elaenia) [3],
Schiffornis turdinus (Thrushlike Manakin) [4],
Pachyramphus polychopterus (White-winged Becard) [5],
Pipreola arcuata (Barred Fruiteater) [6],
Pipra erythrocephala (Golden-headed Manakin) [7],
Mionectes olivaceus (Olive-striped Flycatcher) [8],
Furnarius rufus (Rufous Hornero) [9],
Dendrocolaptes certhia (Barred Woodcreeper) [10],
Formicarius colma (Rufous-capped Antthrush) [11],

	1	2	3	4	5	6	7	8	9	10	11	12	13	14	15	16
1		4.5 (3)	5.4 (2)	8.7 (1)	7.8 (1)	8.7 (4)	7.6 (2)	9.4 (1)	14.3 (1)	-	-	14.2 (1)	13.0 (1)	13.1 (1)	-	14.5 (1)
2	4.6 (2)		5.3 (1)	10.6 (1)	9.2 (1)	8.7 (3)	8.4 (4)	9.2 (1)	-	-	-	-	-	-	-	-
3	6.5 (2)	5.9 (3)		9.1 (1)	9.0 (1)	8.9 (1)	8.8 (1)	10.0 (3)	-	-	-	-	-	-	-	-
4	-	-	-		6.7 (2)	-	8.5 (1)	11.5 (1)	-		-	12.9 (1)	-	14.3 (1)	-	-
5	-	9.5 (2)	8.7 (1)	7.7 (1)		-	8.3 (1)	9.8 (2)	-		11.8 (1)		-	-	-	-
6	8.9 (1)	9.8 (2)	-	10.2 (1)	10.0 (1)		-	-	14.2 (1)	14.4 (1)	14.5 (1)	13.4 (1)	-	13.2 (1)	-	-
7	9.4 (3)	-	-	9.0 (2)	8.8 (1)	8.4 (3)		10.2 (1)	14.2 (1)	-	-	-	-	14.3 (1)	-	14.9 (1)
8	-	9.7 (2)	9.5 (2)	10.7 (1)	9.9 (1)	-	-		13.9 (1)	-	12.5 (1)	-	-	-	-	14.6 (1)
9	14.1 (1)	-	-	-	-	14.3 (1)	13.6 (1)	-		7.1 (2)	-	-	-	10.5 (1)	-	13.3 (1)
10	14.0 (1)	-	-	-	-	13.5 (1)	13.9 (1)	-	7.0 (2)		-	-	-	9.4 (1)	11.2 (1)	13.6 (1)
11	-	13.7 (1)	-	-	-	-	13.3 (1)	14.4 (1)	12.1 (1)	12.8 (1)		11.3 (3)	11.2 (1)	11.2 (3)	11.3 (1)	14.1 (2)
12	-	-	-	-	-	-	13.6 (1)	-	-	12.4 (1)	-		-	10.2 (1)	-	13.4 (1)
13	-	13.2 (1)	-	-	14.4 (1)	-	13.9 (1)	14.5 (1)	12.4 (1)	-	10.8 (2)	10.9 (1)		1.8 (1)	5.8 (2)	-
14	14.5 (1)	13.4 (1)	-	-	-	13.4 (1)	13.3 (1)	-	12.0 (1)	-	-	10.6 (3)	2.0 (2)		5.8 (1)	13.6 (2)
15	-	14.7 (1)	-	-	14.6 (1)	-	13.8 (1)	14.5 (1)	12.2 (1)	-	11.4 (2)	11.2 (1)	6.1 (1)	6.3 (1)		-
16	13.1 (1)	13.3 (1)	-	-	-	13.0 (1)	12.9 (1)	-	12.9 (1)	-	-	12.5 (1)	12.9 (1)	12.6 (2)	12.8 (1)	

Table 9.1: DNA-DNA hybridization data of Sibley & Ahlquist (1985). $\Delta T_{50}H$ means above number of replicates. The first column indicates the species used as tracer and the first row indicates the species used as driver in the hybridization experiment. The dash (-) denotes missing values.

Conopophaga castaneiceps (Chestnut-crowned Gnateater) [12],
Scytalopus latebricola (Brown-rumped Tapaculo) [13],
Scytalopus femoralis (Rufous-vented Tapaculo) [14],
Liosceles thoracicus (Rusty-belted Tapaculo) [15],
Thamnophilus schistaceus (Black-capped Antshrike) [16].

An analysis of all involved species of Sibley & Ahlquist (1985) implies a high percentage of missing values in the dissimilarity matrix. Moreover an effect of the choice of a species as *tracer* or *driver* may influence the analysis to some extent. Restricting our analysis to the data of species used both as tracer-species and driver-species we reduce these problems.

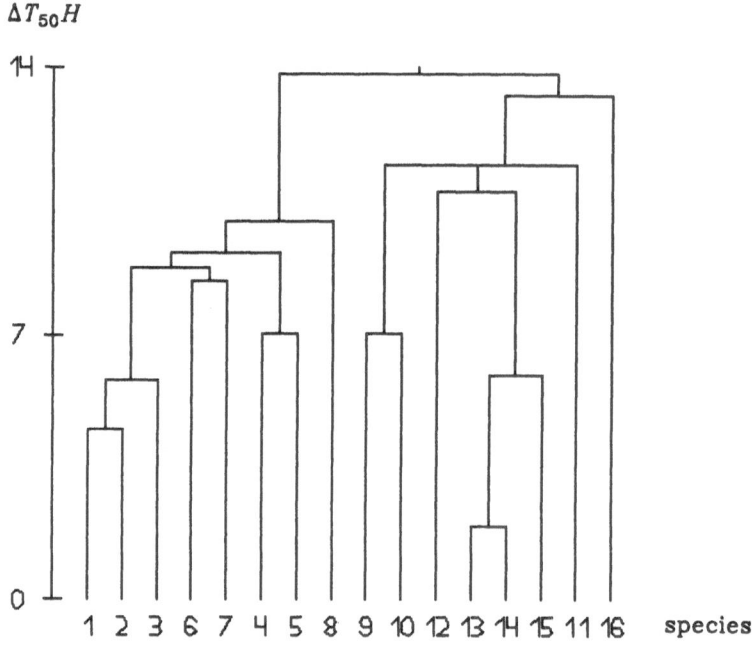

Figure 9.1: The estimated phylogeny of the New World Suboscine Passerine Birds.

The means of the measurements between a tracer-species and a driver-species and the number of replicates are published in Sibley & Ahlquist (1985). Table 9.1 shows the data of Sibley & Ahlquist (1985) between the 16 extracted species. 32 of 120 possible numbers (26.7%) are missing in the resulting dissimilarity matrix. Computing the estimator for the variance of the measurement error we get

$\widehat{\sigma_1^2}$ = 0.5189 by 36 degrees of freedom.

Residual plots gave one hint for a violation of the assumptions about e_1 (cp. (3.2)). The $\Delta T_{50}H$ means between 1 and 7 result in standardized residuals of ±2.737. We have no further information available about single $\Delta T_{50}H$ values, $T_{50}H$ values, sample effects, etc., therefore we do not try to explain hypothetical outliers.

Applying the weighted three objects variance estimator we receive

$$\widehat{\sigma_0^2}_{W3O_b} = 0.1955.$$

Therefore the estimated variance of the measurement error is roughly three times greater than the estimated variance of the random block effect. The estimated ratio of the variance components provides some information for the design of $\Delta T_{50}H$ experiments in the future.

no.	estimated cluster	$\hat{d}_u$	$\hat{l}(\nu(C,A_l))$	$\hat{l}(\nu(C,A_r))$	$\widehat{DP}$
1	1, 2	4.540	-.079	.079	.898
2	4, 5	7.033	.267	-.267	.943
3	6, 7	8.400	.149	-.149	.429
4	9, 10	7.050	.082	-.082	1.000
5	13, 14	1.933	.116	-.116	1.000
6	1 2, 3	5.850	.036	-.036	1.000
7	13 14, 15	5.967	-.224	.211	1.000
8	12, 13 14 15	10.762	-.009	.162	.643
9	1 2 3, 6 7	8.765	-.071	.046	.507
10	1 2 3 6 7, 4 5	9.147	-.075	.124	.879
11	9 10, 12 13 14 15	11.455	.182	-.240	.288
12	11, 9 10 12 13 14 15	11.456	-.104	.210	.999
13	1 2 3 6 7 4 5, 8	9.962	-.101	.154	1.000
14	9 10 11 12 13 14 15, 16	13.235	.064	-.011	.893
15	S	13.799	.081	.012	-

Table 9.2: Stability of estimated phylogeny. Note, estimated cluster $C = A_l, A_r$, where the comma separates the estimated lower neighbours in column two.

The phylogeny of the involved New World Suboscine Passerine Birds is estimated by φ_{AL}^{new} of section 5 using the estimated weights (see section 7). Figure 9.1 shows the estimated phylogeny. We compute the correlation $\hat{\rho}$ (cp. section 8) as three objects method to evaluate the influence of the fluctuations of the evolution. Getting $\hat{\rho} = -0.0794$ we find no evidence for an influence of the fluctuations of the evolution to the data; i.e. we find no evidence for the necessity of the covariance component model (3.2) or a general additive tree model. The residual plots (figure 9.2a,b) indicate two hypothetical outliers, but they imply no other hint for a violation of (3.4) or the necessity of a transformation of the dissimilarity data. The standardized residuals for (10,14) and (5,11) of -3.654 and -3.088 indicate the two hypothetical outliers. The Q-Q-plot (figure 9.2b) represents a correlation of 0.98. The F-statistic of the lack of fit test (cp. section 8) gives a value of 1.106 by (14,59) degrees of freedom and therefore implies no hint for the necessity of an additive tree model. The residual plots under the additive tree design (figure 9.2c,d) indicate the same two outliers.

In order to evaluate the estimated clusters we compute the bootstrap measure of fit $\widehat{DP}$ for estimated clusters. $\widehat{DP}$ is computed by a Monte Carlo study with 1000

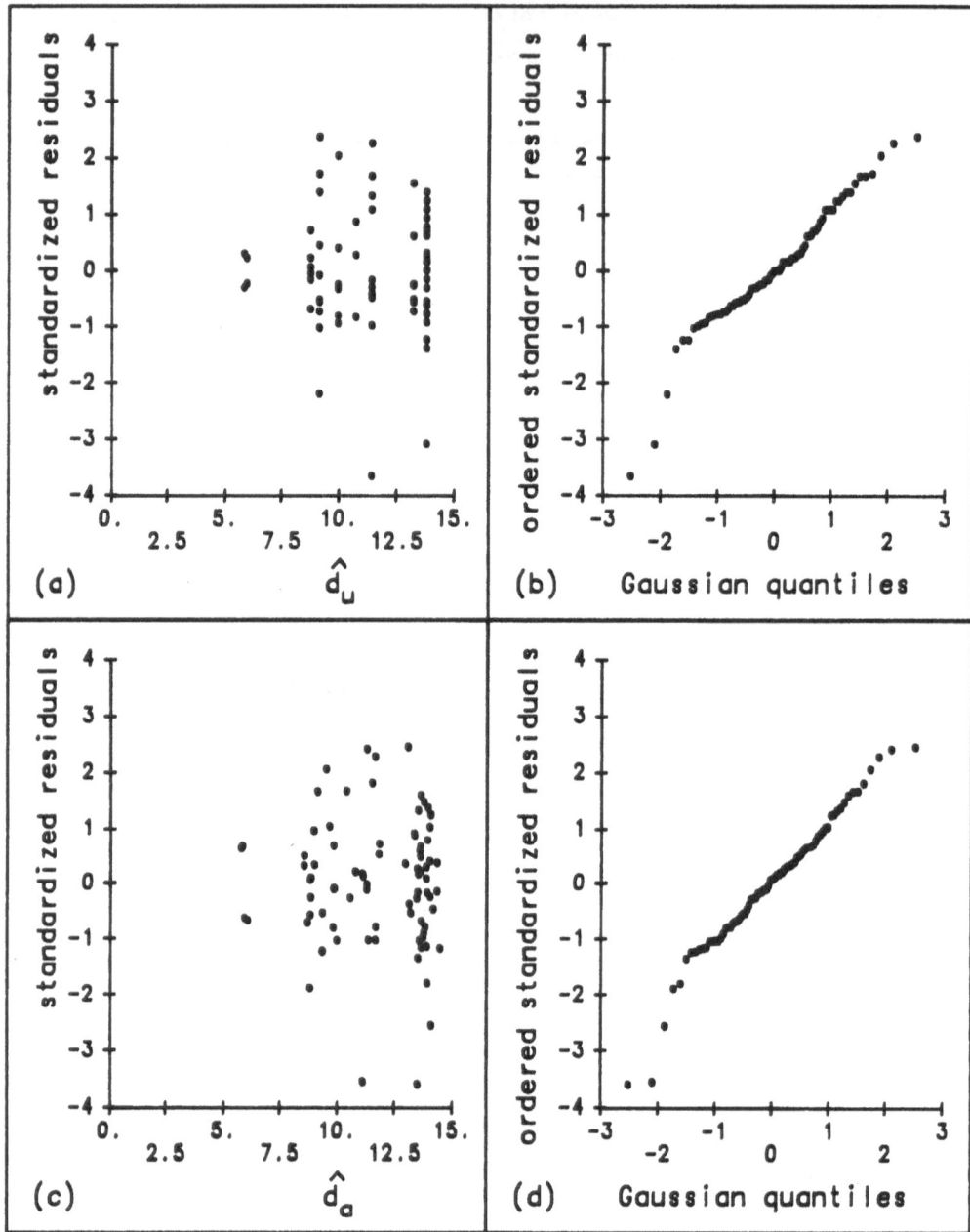

Figure 9.2: Residual analysis for linear models induced by the φ_{AL}^{new} estimate of the hierarchy. (a) and (b): ultrametric design; (c) and (d): additive tree design.

repetitions. Table 9.2 contains the estimated clusters, the cluster levels, the weighted least square estimate for the random fluctuations of the evolution (cp. model (3.2)) and the bootstrap measure of fit $\widehat{DP}$.

Having analysed the data and especially noting the computed bootstrap measures of fit, we have statistical evidence for the estimated clusters no. 6, no. 7, no. 12 and no. 13 by $\widehat{DP} \geq 0.99$. The species of cluster no. 12 are the involved species of the Parvorder *Furnariida* suggested by Sibley & Ahlquist (1985), the species of no.13 are the involved species of the suggested Parvorder *Tyrannida*, the species of no.7 are the involved species of the suggested Family *Rhinocryptidae* and the species of no.6 are the involved species of the suggested Subfamily *Tyranninae*. Consequently the proposed evaluation procedure of the DNA-DNA hybridization data of the New World Suboscine Passerine Birds supports the analysed part of the classification suggested by Sibley & Ahlquist (1985).

References

Baker, F.B. (1974) *Stability of two Hierarchical Grouping Techniques, Case I: Sensitivity to Data Errors*. Journal of the American Statistical Association, 69, 440-445

Bock, H.H. (1974) *Automatische Klassifikation*. Vandenhoek & Ruprecht, Göttingen

Buneman, P. (1971) *The Recovery of Trees from Measures of Dissimilarity*. In: Hodson, F.R., Kendall, D.G. and Tautu, P. (eds) *Mathematics in the Archaeological and Historical Sciences*. Edinburgh University Press, 387-395

Carroll, J.D. and Pruzansky, S. (1980) *Discrete and Hybrid Scaling Methods*. In: Lantermann, E.D. and Feger, H. (eds) *Similartity and Choice*. Huber, Bern, 108-139

Cook, R.D. and Weisberg, S. (1982) *Residuals and Influence in Regression*. Chapman and Hall, New York

Degens, P.O. (1983a) *Hierarchical Cluster Methods as Maximum Likelihood Estimators*. In: Felsenstein, J. (ed) *Numerical Taxonomy*. Springer, Berlin, 249-253

Degens, P.O. (1983b) *Hierarchische Clusteranalyse. Approximation und Agglomeration*. In: Dahlberg, I. and Schader, M. (eds) *Automatisierung in der Klassifikation*. Studien zur Klassifikation, 13, INDEKS Verlag, Frankfurt, 189-202

Degens, P.O. (1985) *Ultrametric Approximation to Distances*. Computational Statistics Quarterly, 2(1), 93-101

Degens, P.O., Lausen, B. (1986) *Statistical Analysis of the Reconstruction of Phylogenies With DNA-DNA-Hybridization Data*. Research Report, Department of Statistics, Univesity Dortmund

Degens, P.O. (1988) *Reconstruction of Phylogenies by Weighted Genetic Distances*. In: Bock, H.H. (ed) *Classification and Related Methods of Data Analysis*. Proceedings of the First Conference of the International Federation of Classification Societies (IFCS). North Holland, Amsterdam, 727-739

De Soete, G. (1984) *A Least Squares Algorithm for Fitting an Ultrametric Tree to a Dissimilarity Matrix*. Pattern Recognition Letters, 2, 133-137

Dobson, A.J.,Kruskal, J.B.,Sankoff, D. and Savage, L.J. (1972) *The Mathematics of Glottochronology revisited*. Anthropological Linguistics, 14, 6

Dobson, A.J. (1974) *Unrooted Trees for Numerical Taxonomy*. Journal of Applied Probability, 11, 32-42

Dubes, R., Jain, A.K. (1979) *Validity Studies in Clustering Methodologies*. Pattern Recognition, 11, 235-254

Efron, B. (1982) *The Jacknife, the Bootstrap and Other Resampling Plans*. OBMSF-MSF Regional Conference Series in Applied Mathematics, Philadelphia, USA

Farris, J.S. (1985) *Distance Data revisited*. Cladistics, 1, 67-85

Felsenstein, J. (1983) *Statistical Inference of Phylogenies*. Journal of the Royal Statistical Society, Series A, 146, 246-272

Felsenstein, J. (1985) *Confidence Limits on Phylogenies: An Approach Using the Bootstrap*. Evolution, 39, 783-791

Felsenstein, J. (1986) *Distance Methods: Reply to Farris*. Cladistics 2

Felsenstein, J. (1988) *Estimation of Hominoid Phylogeny From a DNA Hybridization Data Set*. To appear in Journal of Molecular Evolution

Ferguson, A. (1980) *Biochemical Systematics and Evolution*. Blackie, Glasgow and London

Hartigan J.A. (1967) *Representation of Similarity Matrices by Trees*. Journal of the American Statistical Association, 62, 1140-1158

Hartung, J. (1981) *Non-Negative Minimum Biased Invariant Estimation in Variance Component Models*. Annals of Statistics, 9, 278-292

Jardine, C.J., Jardine, N. and Sibson, R. (1967) *The Structure and Construction of Taxonomic Hierarchies*. Mathematical Biosciences, 1, 173-179

Jardine, N. and Sibson, R. (1971) *Mathematical Taxonomy*. Wiley, New York

Johnson, S.C. (1967) *Hierarchical Clustering Schemes*. Psychometrika, 32, 241-254

Krivanek, M. (1986) *On the Computational Complexity of Clustering*. In: Diday, E. et al. (eds) *Data Analysis and Informatics, IV*. North Holland, Amsterdam, 89-96

Lausen, B. (1987) *Zur Beurteilung der Rekonstruktion phylogenetischer Stammbäume anhand genetischer Distanzen*. Diplomarbeit, Fachbereich Statistik, Universität Dortmund

Lausen, B. (1988) *Maximum Likelihood Agglomeration in Hierarchical Cluster Analysis*. To appear in: Wille, R. (ed) *Klassifikation und Ordnung*. Studien zur Klassifikation, 19, INDEKS Verlag, Frankfurt

Lausen, B. and Degens, P.O. (1986) *Variance Estimation and the Reconstruction of Phylogenies*. In: Degens, P.O., Hermes, H.-J. and Opitz, O. (eds) *Classification and its Environment*, Studien zur Klassifikation, 17, INDEKS Verlag, Frankfurt, 306-314

Lausen, B. and Degens, P.O. (1987) *Bootstrap Evaluation in Hierarchical Cluster Analysis*. Proceedings of the 5th International Symposium on Data Analysis and Informatics at Versailles. Institut National de Recherche en Informatique et en Automatique (INRIA), 267-276

Lausen, B. and Degens, P.O. (1988) *Evaluation of the Reconstruction of Phylogenies With DNA-DNA-Hybridization Data*. In: Bock, H.H. (ed) *Classification and Related Methods of Data Analysis*. Proceedings of the First Conference of the International Federation of Classification Societies (IFCS), North Holland, Amsterdam, 367-374

Lausen, B. and Vach, W. (1986) *Estimation, Graphical Representation and Judgement of Evolutionary Trees in Expert Systems*. In: Haux, R. (ed) *Expert Systems in Statistics*. Fischer, Stuttgart, 61-74

Ostermann, R. and Degens, P.O. (1984) *Eigenschaften des Average-Linkage-Verfahrens anhand einer Monte-Carlo-Studie*. In: Bock, H.H. (ed) *Anwendungen der Klassifikation: Datenanalyse und numerische Klassifikation*. Studien zur Klassifikation, 15, INDEKS Verlag, Frankfurt, 108-114

Ostermann, R. and Degens, P.O. (1986) *Die Qualität des Average-Linkage-Verfahrens bei verschiedenen Fehlerverteilungen*. EDV in Medizin und Biologie, 16(3), 91-96

Rothe, G. (1988) *Bootstrap: Estimating and Testing*. Statistical Software Newsletter, 14, 31-34

Sibley, C.G. and Ahlquist, J.E. (1983) *Phylogeny and Classification of Birds Based on the Data of DNA-DNA Hybridization*. Current Ornithology, 1, 245-292

Sibley, C.G. and Ahlquist, J.E. (1984) *The Phylogeny of the Hominoid Primates, as Indicated by DNA-DNA Hybridization*. Journal of Molecular Evolution, 20, 2-15

Sibley, C.G. and Ahlquist, J.E. (1985) *Phylogeny and Classification of New World Suboscine Passerine Birds (Passeriformes: Oligomyodi: Tyrannides)*. American Ornithology Union Monograph, 36, 396-428

Simoes Pereira, J.M.S. (1969) *A Note on the Tree Realizability of a Distance Matrix*. Journal of Combinatorial Theory, 6, 303-310

Sneath, P.H.A. and Sokal, R.R. (1973) *Numerical Taxonomy*. W.H. Freeman, San Francisco

Vach, W. and Degens, P.O. (1986) *Starting More Robust Estimation of Ultrametrics*. In: Degens, P.O., Hermes, H.-J. and Opitz, O. (eds) *Classification and its Environment*. Studien zur Klassifikation, 17, INDEKS Verlag, Frankfurt, 239-246

Vach, W. and Degens, P.O. (1988a) *The System of Common Lower Neighbours of a Hierarchy*. In: Bock, H.H. (ed) *Classification and Related Methods of Data Analysis*. Proceedings of the First Conference of the International Federation of Classification Societies (IFCS), North Holland, Amsterdam, 165-172

Vach, W. and Degens, P.O. (1988b) *Improvement of the Average Linkage Method*. Research Report, Department of Statistics, University Dortmund

Vach, W. and Degens, P.O. (1988c) *A new Average Linkage Method*. To appear in: Wille, R. (ed) *Klassifikation und Ordnung*. Studien zur Klassifikation, 19, INDEKS Verlag, Frankfurt

Weir, B.S. (1988) *Statistical Analysis of DNA Sequences*. Journal of the National Cancer Institute, 80, 395-406

Additive-Tree Representations

Hervé Abdi[1]

University of Bourgogne at Dijon

PURPOSE OF ADDITIVE-TREE REPRESENTATIONS

Additive-trees are used to represent objects as "leaves" on a tree, so that the distance on the tree between two leaves reflects the similarity between the objects. Formally, an observed similarity δ is represented by a tree-distance d. As such, additive-trees belong to the descriptive multivariate statistic tradition. Additive-tree representations are useful in a wide variety of domains. For example:

Filiation of Manuscripts (Buneman, 1971)

This application was suggested (but not actually performed) by Buneman (1971) in a seminal paper that laid the foundations for additive-tree representations. Here, the objects to be represented were manuscript copies of the same text (e.g., as in the medieval tradition). The problem is to infer from the set of texts the *family tree* that generated the variants (i.e., which manuscript is copied from which other one; are there lost copies? etc.). The similarity between texts can be defined, for example, as the number of *common errors*, or more simply as the number of common words. Examples of filiation data are given in the volume edited by Hodson, Kendall & Tartu (1971) in which the Buneman paper has been published. Another example of this line of investigations can be found in Abdi (1985, 1989b).

---

[1] Correspondence about this paper should be adressed to: Hervé Abdi, Université de Bourgogne, Laboratoire de Psychologie, Ancienne Faculté 36 rue Chabot-Charny, F-21000 Dijon, France. The author wishes to thank Sue Viscuso and Alice O'Toole for help and comments on previous draft.

Psychological similarity between animal terms

Henley (1969) conducted several experiments on semantic memory structure. She obtained from 21 subjects an estimation of the "subjective distance" between 12 animal terms. Subjects were asked to list from memory the animal names they knew. For each animal pair the number of animals separating the pair was divided by the list length. For example, suppose that the "cow-mouse" pair was separated by 7 animals, and that the total number of animals given was 20, then the value attached to the pair "cow-mouse" will be 7/20 = .35. This procedure was repeated for each subject. Then, the value for each pair was collapsed across subjects to obtain the average similarity. Abdi, Barthélemy & Luong (1984) used a tree (percentage of explained variance: 73) to represent this data matrix. The tree is displayed in the first Figure.

Phylogenetic trees

The literature concerning phylogenetic trees is very large; some influential papers are Phipps (1971); Farris (1973); Waterman, Smith, Singh & Beyer (1977). Biologists have often tried to describe the relation between contemporary species. The similarity between species is defined, for example, as the number of identical sequences for some protein or for the DNA, etc. The leaves represent the actual species, and (as in the filiation problem of Buneman) the interior vertices can be interpreted as "missing links" or common ancestors. In general, biologists focus their interest more on the *shape* of the tree rather than on the *distance* between vertices of the tree, because it is more important in this context to assess the existence of common ancestors for some species rather than to suggest when the separation of the species did occur. An example of phylogenetic tree is given in Figure 2 (cf. Dress & Krüger, 1987). Because different similarity measures can suggest different patterns of evolution, one problem faced by biologists is to compare different trees obtained from a same set of objects. A related problem is to define a tree that expresses the consensus among different trees (cf. Bobisud & Bobisud; Robinson, 1971; Robinson & Foulds, 1981).

BASIC TREE-NOTIONS

Recall that distances between pairs is a function denoted by d, that associates a positive real value to each pair of a given set, satisfying the following conditions:

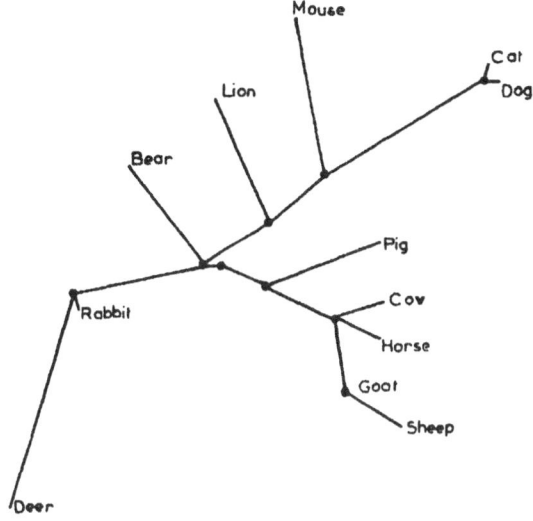

Figure 1. : Additive-tree representation of Henley's data (1969), from Abdi *et al* (1984).

$$d_{i,j} \geq 0$$
$$d_{i,i} = 0$$
$$d_{i,j} = d_{j,i}$$
$$d_{i,j} \leq d_{i,k} + d_{k,j} \quad \text{(triangle inequality)}$$

Four-points condition

A distance is a tree distance (i.e., derived from a tree) if and only if the so called *four-points condition* is satisfied. This condition expresses that four points on a tree can always be labeled x, y, z, t such that:

$$d_{x,y} + d_{z,t} \leq (d_{x,z} + d_{y,t}) = (d_{x,t} + d_{y,z})$$

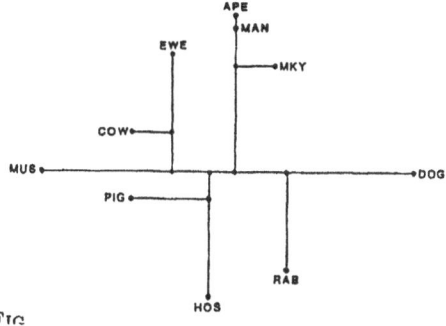

FIG

Figure 2. : a phylogenetic tree from Dress & Krüger (1987).

This condition has been discovered by several authors in different contexts (Zarestkii, 1965; Simões-Pereira, 1969; Buneman, 1971; Patrinos & Hakimi, 1972; Dobson, 1974; etc.).

Note: the four-points condition clearly implies the triangle inequality. Intuitively, it seems clear that the four-points condition in turn is implied by the ultrametric inequality (for a proof, see Dobson, 1974). The ultrametric inequality is expressed as:

$$d_{x,y} \leq \max\{d_{x,z}, d_{z,y}\} \quad \forall x,y,z$$

Thus, ultrametric trees represent a particular case of additive trees.

Strict and weak score, split and H-relations

The four-points condition can be expressed as a relation between pairs of vertices. An essential tool for understanding trees is the notion of *score* of a pair of vertices. When d is a tree-distance, then all 4-vertices can be labelled x,y,z,t such that either (a) or (b) holds:

$$(a) \quad (d_{x,y} + d_{z,t}) < (d_{x,t} + d_{y,z}) = (d_{x,z} + d_{y,t})$$
$$(b) \quad (d_{x,y} + d_{z,t}) = (d_{x,t} + d_{y,z}) = (d_{x,z} + d_{y,t})$$

Following Colonius & Schulze (1981), condition (a) is denoted $\mathbf{xyH^x zt}$ and condition (a) or (b) is denoted $\mathbf{xyHzt}$. An illustration of the conditions (and of their) names is given in Figure 3.

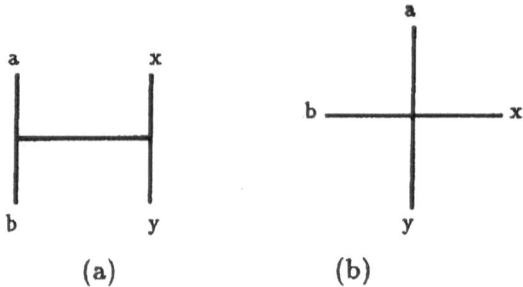

(a)   (b)

Figure 3. : The two configurations of 4 points on a tree, the H and the H^x relations.

The H-relation can be expressed in different ways. When $\mathbf{xyH^x zt}$, the pair (x,y) is said to be *split* from the pair (z,t) (Buneman, 1971). In terms of tree structure, this means that by deleting one edge of the tree, two non-connected sub-trees are created. One tree contains the vertices x,y, the other one contains the vertices z,t. For example, if in Figure 3.a the internal edge is deleted, two trees will be created, the first one consisting of x,y, the second one consisting of a,b. These conditions can also be expressed in terms of a quaternary relation (i.e., a binary relation between pairs). Following Buneman (1971) and (Dobson, 1974), Colonius & Schulze (1981) developed this idea and characterized tree-distances from the (non-numerical) properties of the H^x relation as follows: for all x, y, z, v, t

- *i.* $\textbf{xy}H^x tz$ implies $tz H^x xy$ and $yx H^x tz$.
- *ii.* $xy H^x tz$ implies neither $xz H^x ty$ nor $tx H^x yz$ holds.
- *iii.* $xy H^x tz$ implies either $xy H^x vz$ or $vx H^x tz$ or both.
- *iv.* $xy H^x tz$ and $xy H^x tv$ implies $xy H^x vz$

Actually, *iv* is implied by *i* , *ii*, and *iii* (Bandelt & Dress, 1986; see also Barthélemy & Luong, 1987).

The *strict score* of the pair (x,y) is the number of pairs (t,z) such that $\textbf{xy}H^x tz$. The *loose score* of the pair (x,y) is the number of pairs (t,z) such that $xy H tz$.

When d is a tree-distance, the tree structure can be recovered from the score matrix (cf. Buneman, 1971; Colonius & Schulze, 1981; Bandelt & Dress, 1986). This fundamental relation can be expressed in topological terms (i.e., in terms of "neighborness" or "neighborliness", cf. Bandelt & Dress, 1986; Barthélemy & Luong, 1987). In a sense, $xy H^x tz$ expresses simultaneously that x and y are neighbors and that x,y are separated from z,t. In other words, x and y are farther from z and t than they are from each other. This notion is generalized by Barthélemy & Luong (1987) and by Barthélemy & Guénoche (1987). From that notion, Eastbrook, McMorris & Meacham (1985) and Day (1985) derive some indices used to compare trees, although some problems may occur when this neighboring relation is used within this framework (McMorris, 1985).

HISTORICAL SKETCH

Interest in additive-tree representations originated in several fields such as operational research, computer sciences, biology and psychology. The contribution of these fields is detailed in the following sections.

Operational research and computer sciences

Perhaps the first proof of the four-points condition is to be attributed to Zarestkii (1965). The condition is proved for a tree with a unitary valuation (i.e., all the edges have the same length). Smolenski (1963) proved the unicity of the tree associated with (unitary) tree-distance. However, because these papers were published in eastern journals, they were not widely known or accessible. Actually, the paper by Zarestkii, written in Russian, was known *via* a review of graph theory in the USSR (Turner & Kautz, 1970). Within this tradition, Boesch (1968), Simões-Pereira (1969) and later Patrinos & Hakimi (1972) characterized the tree-distance and, in the latter reference, characterized the four-points condition for the general case (i.e., "non-unitary"). These authors, incidentally, did not mention Buneman (1971) but referred to Zarestkii (1965) and Simões-Pereira (1969), who in turn were not mentioned by Buneman (1971). In the meantime, tree-metrics have made their way into mathematics proper (cf. Dress, 1984; and the literature quoted there).

Biology

Buneman (1971) in his classic paper recalled that some authors (biologists, although not mentioned explicitly as such) have previously dealt with problems similar to the manuscript filiation problem. Specifically, he referred to work on phylogenetic trees and signaled that Cavalli-Forza & Edwards (1967) and Ecks & Dayhoff (1966) had described methods for finding an approximation of dissimilarity data by a tree distance. Fruitful research is still conducted in this field (cf. among other, Waterman, Smith, Singh & Beyer, 1978; Fitch, 1981; Meacham, 1987; Peacock, 1981).

As mentioned previously, a main problem in biology is to integrate the results of different analyses. That is, to compare either different tree distances, or different trees obtained for one set of objects with different measures, methods or (dis)similarity indices. This problem led Phipps (1971) and Farris (1973) to define the tree-distance when the valuation of the edges is unitary (cf. previous section on Zarestkii). The problem of consensus between different trees, and the definition of a distance between trees is explicitly dealt with by Bobisud & Bobisud (1972). This tradition has been carried on by different authors (Robinson, 1971; Dobson, 1975; Robinson & Foulds, 1981). Currently, the problem of defining a consensus between trees remains open as different approaches can be proposed (cf. Day, 1985; Estabrook *et al.*, 1985; McMorris, 1985; Dress & Krüger, 1987).

The comparison of the approach taken in biology with the approach taken in psychology emphasizes the *rôle* of the field of application in the theoretical developments. That is, for biologists, the shape of the tree is of prime importance while for psychologists the tree distance between leaves is the main concern (cf. the section on tree analysis as regression). As a consequence of this difference of emphasis, the very idea of *goodness of fit* differs among pratictioners. To caricature, for biologists a good tree recovers the shape of a possible "original tree," whereas for psychologists, a good tree preserves the original dissimilarity. Thus biologists favor measures of fit expressed in terms of quaternary relations (e.g., number of violations of the "natural" H-relation by the tree). Psychologists, on the other hand, prefer measures like stress (due to the nonmetric multidimensional scaling tradition), or more adequately, the part of the variance of the original dissimilarity explained by the tree model (i.e., the squared correlation coefficient between the original distance and the tree-distance approximation). Hence, the notion of goodness of fit is far from universal in the tree world...

The idea of emphasizing a topological point of view in interpreting a tree is clearly akin to the biological point of view. However this idea was first proposed by mathematical psychologists Colonius & Schultz (1981) who placed themselves in the measurement theory tradition. This perspective has been expanded recently by Bandelt & Dress (1986) and Barthélemy & Luong (1987).

Psychology

In the psychological tradition, besides Buneman (1971), two papers (Carroll & Chang, 1973;

Cunningham, 1974) have had a strong influence. As an oddity[2] these papers were actually "underground" papers. Precisely, they were given as talks in scientific meetings and never circulated (stricly speaking) as papers. However, these two papers aroused the interest of mathematical psychologists. Eventually, one of these papers was published as part of a broad paper that also explored "directed trees" (Cunningham, 1978).

Almost from the beginning of the interest for additive trees within psychology, two traditions were created. The first, following Carroll & Chang, 1973, comes from the "scaling" or psychometric tradition and considers a tree as a more or less practical *graphical* representation of (dis)similarity data. Consequently, the additive tree representations are contrasted with the more usual maps derived by nonmetric multidimensional scaling (cf. the review of Shepard, 1974, 1980; Carroll & Arabie, 1980). Here, the main problem is to decide when to favor tree representations over euclidean maps. Different criteria have been proposed, but stress or percentage of explained variance (i.e., square correlation between the original dissimilarity and the tree distance) are generally preferred. Pruzansky, Tversky & Carroll (1982) offer some guidelines that can help to decide, based on data properties (i.e., dispersion, skweness, proportion of elongated triangles, etc.), which representation is more appropriate for a given set of data. The main result of these investigations is that, on the whole, Euclidean distance data show positive skewness, and tree distance data show negative skewness.

The second tradition is linked to cognitive psychology, specifically to work in semantic memory organization. Precisely, the empirical work of Rosch initiated this work (for reviews, see, for example, Rosch, 1975, 1983; Abdi, 1986a). Tversky (1977, Tversky & Gati, 1977) proposed a formalization of the notion of "psychological similarity" in parallel with his work on additive tree fitting with Sattath (Sattath & Tversky, 1977).

The theory was designed to give an account of several empirical observations, including the *typicality effect*. This term is used by psychologists to express that some members of a natural category are more representative of that category than are some of the other members. The classic example is that, for the category *bird*, canary is a better representative of the category than is penguin. As such, class elements are not equivalent and are often more or less representative of their category. Clearly, the standard ultrametric tree representation forces objects to be equivalent (i.e., in an ultrametric tree, all the elements of a class are at the same distance from the center of the class). As such these trees do not adequately represent data that have a gradient of representativity. The additive tree, with its edges of varying lengths, seems more appropriate. Tversky (1977) went further than simply advocating tree representations as a tool. He showed that the additive tree distance is a particular case of a more general model of similarity named the *contrast model*. This theme is expanded further in this paper in the section on tree interpretation.

In fact, semantic memory has been a fruitful source of inspiration for psychologists. For example, Schulze & Colonius (1979) and Colonius & Schulze (1981) were originally interested in the exploration of the semantic meaning of verbs. To do so, they used directly the quaternary H^x-relation and asked subjects to group quadruplets of verbs in pairs (cf. the IVb quadrant of Coombs's *theory of data*, 1964). This task gave the impetus for their later characterization of the tree distance in terms of the H^x-relation.

---

[2] that would probably please people at the institute for scientific information.

ALGORITHMS FOR ADDITIVE TREE APPROXIMATION

The algorithms for tree approximation can be roughly divided into two large families.

Algorithms of the first family proceed by estimating the tree structure, and then adjusting for the length of the edges in order to fit the "original distance". The structure of the tree is determined, in general, using the score matrix (e.g., Sattath & Tversky, 1977). The score matrix can be recomputed after each iteration (as in Sattath & Tversky), or approximated (for example by using a "quasi single link" method on the score matrix as in Abdi *et al*, 1984; Barthélemy & Luong, 1985; Abdi, 1989a). The edge length can be evaluated either by the least squares method or by diverse geometrical means.

The algorithms of the second family (Cunningham, 1978; Carroll & Pruzansky, 1980; De Soete, 1980; Roux, 1986; Brossier, 1987) proceed by trying to find a best tree distance (in some pre-defined sense) approximating the original distance or dissimilarity and then by constructing the tree associated with the tree distance. This second step is indeed straightforward, but the first step is strongly dependent on the criterion choosen to evaluate what is meant by "best approximation". Cunnigham (1978) and De Soete (1980) use a least squares criterion. Alternatively, Roux (1986) minimizes the number of quadruplets that disagree with condition (*a*) of section II. Following suggestions from Carroll (1976), several authors (Carroll & Pruzansky, 1980; Carroll, Clark & DeSarbo, 1985; Brossier, 1986; Brossier & Calvé, 1986) use the property that an additive tree distance is decomposable into the sum of an ultrametric distance and a distance to center (i.e., a "star distance"). These authors estimate separately the two components, and reconstitute the tree distance by the sum of its two components.

A detailed description of tree approximation algorithms can be found in Luong (1987), along with comparisons relative to complexity, accuracy, etc. Some comparisons on small data sets ($n = 7$) can be found in Guénoche (1987). Finally, a general framework is given in Barthélemy & Guénoche (1988).

HOW TO ANALYZE A TREE THE "TVERSKY WAY"

[3] Tversky (1977) developed a general approach to similarity called the *contrast model*. Each stimulus (say, *a*,*b*, ...) is associated with a set of features (denoted *A*,*B*, ...). The similarity from *a* to *b* is defined as a function of the three sets: $A \cap B$, A/B, and B/A.

In sum:

$$s(a,b) = F(A \cap B, A/B, B/A)$$

If some additional conditions are imposed (namely: monotonicity, independence, along with the two "technical conditions" of solvability and invariance),[4] then the similarity between *a* and *b* can be expressed as:

$$S(a,b) = \alpha f(A \cap B) - \beta f(A/B) - \gamma f(B/A)$$

---

[3] This section and the following are adapted from Abdi, 1986b

[4] Actually, it is possible to derive the contrast model from a slightly different set of axioms, as shown by Osherson (1987).

where f is an interval scale function (cf. Krantz *et al.*, 1971), and α, β, γ are positive constants.

Moreover, when $\beta = \gamma$ (which is necessary for symmetry), and f is additive (i.e., $A \cup B = f(A) + f(B) - f(A \cap B)$), then there exists a measure g such that (cf. Sattath, 1976):

$$S(a,b) = \lambda - g(A/B) - g(B/A) = \lambda - g(A \Delta B)$$

with λ a positive constant, and $A \Delta B$ being the set symmetric difference between A and B (i.e., the number of features that belong to either a or b, but not to both). Now, if the features follow a tree model (that is if three stimuli can always be named a, b, c with $A \cap B = A \cap C \subset B \cap C$), this property allows a very convenient tree representation of the similarities. If the stimuli are leaves on the tree, and the edges are appropriately valued, then the tree distance from a to b will be

$$d(a,b) = g(A/B) + g(B/A) = g(A \Delta B).$$

This is equivalent to defining the distance in terms of distinctive features.

This expression of the distance on the tree in terms of distinctive features can be used to estimate the features composing the stimuli. To see how this is done, it is convenient to introduce three notions:

- the *median* of a tree.
- the *eccentricity* of a vertex (eccentricity is to be taken here as "distance from a center" according to its etymology).
- the *intersection vertex* of two vertices.

The *median* of a tree is the vertex that minimizes the sum of the distances to the set of the vertices. The *eccentricity* of vertex a, denoted $e(a)$ is defined as the distance from a to the median of the tree. The *intersection vertex* of vertices a and b is the vertex with minimal eccentricity situated on the path from a to b. The reason for this naming will become obvious later on. Call a, b two vertices and x their intersection vertex, then the set components of the contrast model are obtained by:

$$\begin{aligned} g(A/B) &= d_{a,x} = e(a) - e(x) \\ g(B/A) &= d_{b,x} = e(b) - e(x) \\ g(A \cap B) &= e(x) \end{aligned}$$

Note, incidentally, that $e(a) = g(A)$ can be seen as a measure of the overall saliency of stimulus a.

Thus, the additive tree representation can be used as a tool to recover *a posteriori* the features (and their weights) composing a set of stimuli from a distance matrix. In particular, a tree can give the number of distinctive features common to every pair of stimuli, and decompose each stimulus into (weighted) features so that the distance on the tree between stimuli is computed simply as a distance between (weighted) features. To do so, it is sufficient to use the "city-block distance", which can be interpreted as a generalization of the symmetric difference distance (for more details see Abdi, 1985). In this sense, a additive-tree analysis can be seen as a regression model, or a factorial method. All these notions, hopefully will be made clear by a example.

Portraits	A	B	C	D	E	F	G	H	I	J	K	L	M	N	O	P	Q	R	S	T
Mass murderer	11	6	3	2	0	0	0	18	2	8	5	1	1	0	0	1	1	0	0	1
Armed Robber	3	4	0	4	0	4	0	13	0	8	4	1	7	1	0	3	0	5	2	1
Rapist	5	2	1	7	4	5	0	4	1	18	0	0	1	1	0	8	0	1	0	0
Medical Doctor	0	1	2	0	2	1	12	1	0	0	8	8	0	2	4	2	5	2	4	6
Clergyman	0	1	5	0	1	1	8	0	2	0	4	1	1	0	6	2	20	1	3	4
Engineer	0	1	1	0	1	3	8	0	1	0	1	5	3	4	8	5	6	2	4	9

Table 1. : Association between 20 faces and 6 occupations. At the intersection between one row and one column the number of subjects that select the occupation for the face is displayed. (data from Goldstein *et al.*, 1984).

GOOD GUYS AND BAD GUYS: A TREE ANALYSIS

In a recent paper, Goldstein, Chance & Gilbert (1984) made a new contribution to the topic of "implicit physiognomy." Implicit physiognomy means that observers agree among themselves and find it meaningful to attribute personality traits, intentions, occupations, etc. merely by looking at a face or a photograph of a face. In the Goldstein *et al.* study, subjects were presented with five arrays each composed of 20 photographs of white middle-aged men taken from a casting directory, and asked to find among the 20 portraits of each array three bad guys (mass murderer, armed robber, rapist) and three good guys (medical doctor, clergyman, engineer). The data were analyzed *via* 30 chi-square tests (one for each of the six "occupations" of the 5 arrays). Because 27 of these 30 tests were statistically significant at the .05 level, it was concluded that there was indeed a clear consensual agreement among the subjects. As an illustration, Goldstein *et al.* gave a contingency table corresponding to the results obtained from 58 subjects with the third array they used (i.e., they gave the number of subjects that assigned a given occupation to each portrait from array #3). These results are summarized below. Only the analysis on the occupations is reported here. The analysis proceeds in two steps. First a distance matrix between occupations will be computed. Second, a tree will be computed to fit these data.

	Doctor	Engineer	Clergyman	Murderer	Robber	Rapist
Doctor	0	34	46	90	84	96
Engineer		0	44	98	80	86
Clergyman			0	86	88	98
Murderer				0	44	70
Robber					0	58
Rapists						0

Table 2. : City-block distance matrix computed from Table 1

Distance matrix between occupations

The strong connection between the tree distance and the city block metric justifies using the city block metric to compute distance between occupations. Precisely, if $k_{i,j}$ denotes the number of subjects that assigned face i to occupation j, then the distance between two occupations j and j' is computed by:

$$d_{j,j'} = \sum_i |k_{i,j} - k_{i,j'}|$$

	Doctor	Engineer	Clergyman	Murderer	Robber	Rapist
Doctor	0	34.00	46.00	91.25	82.75	94.00
Engineer		0	44.00	89.25	80.75	92.00
Clergyman			0	93.25	84.75	96.00
Murderer				0	44.00	68.00
Robber					0	59.25
Rapists						0

Table 3. : Tree-distance approximation of distance matrix from Table 2

Table 2 gives the city block distance matrix, Table 3 gives the tree distance matrix, and figure 4 display the tree. The squared correlation coefficient between the original matrix and the tree distance matrix is .972. It is denoted by the letter τ, and is taken as an index of goodness of fit. Its value indicates a fairly good fit between the data and a tree-model.

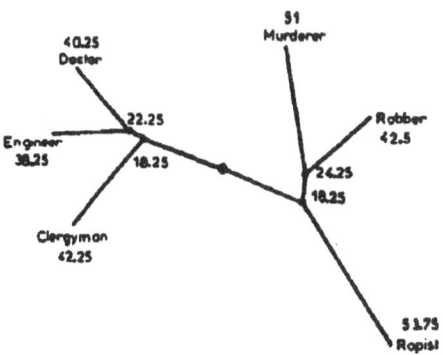

Figure 4. : Additive tree of the occupation distance matrix from Table 2. The number near each vertex is the eccentricity of the vertex.

Names of the features	Occupations					
	Doctor	Engineer	Clergyman	Murderer	Robber	Rapist
Doctor	18.00					
Engineer		16.00				
Clergyman			24.00			
Murderer				26.25		
Robber					17.75	
Rapists						35.50
Engl. & Doct.	4.00	4.00				
Engl. & Doct. & Cler.	18.25	18.25	18.25			
Murd. & Robb.					6.50	6.50
Murd. & Robb. & Rapi.				18.25	18.25	18.25
Sum = eccentricity	40.25	38.25	42.25	51.00	42.50	53.75

Table 4. : Tree-reconstitution of the specific and common features

	Doctor	Engineer	Clergyman	Murderer	Robber	Rapist	Sum
Doctor		22.25	18.28	0	0	0	40.50
Engineer	22.25		18.25	0	0	0	40.50
Clergyman	18.25	18.25		0	0	0	36.50
Murderer	0	0	0		24.25	18.25	43.30
Robber	0	0	0	24.75		18.25	43.00
Rapists	0	0	0	18.25	18.25		36.50

Table 5. : Tree-reconstitution of the common features

The tree shows that the subjects clearly separated the "bad guys" from the "good guys." Moreover, the stereotypes of the good guys are less differentiated than the stereotypes of the bad guys. In particular, the rapist is the most stereotyped occupation, while the engineer is the least stereotyped. These conclusions are supported by Table 5 where the eccentricity of the occupations is decomposed into specific and common weighted features. Thus, the tree analysis can be seen as a variety of the classical factorial analysis. As pointed out previously, this "canonical weighted features matrix" is equivalent to the tree distance matrix when the computed distance is the city block distance. Note that the features are labelled in agreement with the contrast model interpretation of the tree.

WHEN TO USE A TREE? DIRECTIONS FOR USE....

Perhaps due to some halo effect, it is sometimes thought that the prescriptions for additive tree representation parallel those for nonmetric multidimensional scaling. It should be emphasized that additive tree representations are meaningful only for data that are invariant by linear transformation (i.e., interval scale measurement, cf. Suppes & Zinnes, 1963; Roberts, 1979).

By contrast, nonmetric dimensional scaling deals with data matrices that are supposed to be invariant by monotonic transformation. An equivalent way of expressing this condition with the measurement theory vocabulary is that the data are supposed to be measured on an ordinal scale.

(a)						(b)			
	x	y	z	t			x	y	z	t
x	0	3	7	8		x	0	6	7	8
y		0	8	9		y		0	8	10
z			0	11		z			0	11

Table 6. : Two dissimilarity matrices. Matrix *b* is obtained by a monotonic transformation of the values of matrix *b*.

An example from De Soete (1983) clearly confirms that additive trees are non meaningful for data measured on a ordinal scale. The two dissimilarity matrices *a* and *b* of Table 6 are equivalent for the set of monotonic transformations (*b* is obtained by changing in *a* 3 to 6 and 9 to 10):

Unfortunately, as shown in Figure 5, the trees obtained from these two matrices are topologically different, which means that tree representations are not invariant for monotonic transformations. As a consequence, the application domain of additive trees is much more restricted than the domain of multidimensional scaling. Actually, it can be shown that additive-tree representations are meaningful only for an *interval scale* measurement (i.e., additive-tree representations are invariant for the set of linear transformations, see Brossier, 1985; or Barthélemy & Guénoche, 1988).

CONCLUSION

Although this introduction to additive tree representations is an incomplete one, hopefully the presentation of examples were sufficient to enable the reader see the forest for the tree. The reference section serves as a guide for readers interesting in learning more about the topic. Some recent developments should also be alluded to here. In particular, the trees described in this paper were built from complete 2-way square matrices. But trees can also be computed from rectangular matrices (Furnas, 1980; Brossier, 1986), three ways matrices (Carroll, Clark & DeSarbo, 1984) or incomplete matrices (De Soete, 1984). An other line of research has been pioneered by Corter & Tversky (1987), who tried to relax the tree constraint in order to represent proximity data by an extended tree. These extended trees generalize traditional trees by including marked segments that correspond to overlapping clusters. Closely related to this approach is the notion of weak hierarchies recently proposed by Bandelt & Dress (1988).

Hence trees are still growing and bearing diverse fruits.

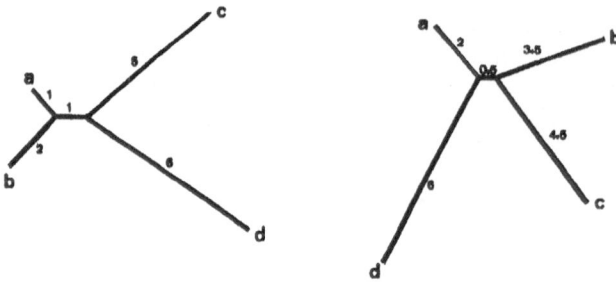

Figure 5. : Additive tree representation of matrix (a) and (b) from Table 6

REFERENCES

Abdi, H. (1985)—Représentations arborées de l'information verbatim. In J., Wittwer (Ed.): *Psycholinguistique Textuelle*. Paris: Bulletin de Psychologie.

Abdi, H. (1986a)—La mémoire sémantique, une fille de l'intelligence artificielle et de la psychologie. In C., Bonnet, J.C., Hoc, G., Tiberghien (Eds.): *Psychologie, Intelligence Artificielle et Automatique*. Bruxelles: Mardaga.

Abdi, H. (1986b)—Faces, prototypes, and additive tree representations. In H.D., Ellis, M.A., Jeeves, F., Newcombe, A., Young (Eds.): *Aspects of Face Processing*. Dordrecht: Nijhoff.

Abdi., H. (1989a)—ABD-TREE: A complete Apple 2 set of programs for additive-tree representations. In X. Luong (Ed.): *Additive-Tree Representations of Textual Data*. Nice: Cumfid (Special Issue).

Abdi., H. (1989b)—Additive-tree representations of verbatim memory. In X. Luong (Ed.): *Additive-Tree Representations of Lexical Data*. Nice: Cumfid (Special Issue).

Abdi., H., Barthélemy, J.P., Luong, X. (1984)—Tree representations of associative structures in semantic and episodic memory research. In E., Degreef, J. Van Huggenhaut (Eds.): *Trends in Mathematical Psychology*. New York: Elsevier.

Bandelt, H.J., Dress, A. (1986)—Reconstructing the shape of a tree from observed dissimilarity data. *Advances in Applied Mathematics*, 7, 309-43.

Bandelt, H.J., Dress, A. (1988)—*Weak hierarchies associated with similarity measures*. Unpublished manuscript. Bielefeld University.

Barthélemy, J.P., Guénoche, A. (1988)—*Arbres et Représentations des Proximités*. Paris: Masson.

Barthélemy, J.P., Luong, X. (1986)—Représentations arborées des mesures de dissimilarité. *Statistiques et Analyse des Données*, 5, 20-41.

Barthélemy, J.P., Luong, X. (1987)—Sur la topologie d'un arbre phylogénétique: aspects théoriques, algorithmes et applications à l'analyse des données textuelles. *Mathématiques et Sciences Humaines*, 100, 57-80.

Bobisud, H.M., Bobisud, A. (1972)—A metric for classification. *Taxonomy*, **21**, 607-613.

Boesch, F.T. (1968)—Properties of the distance matrix of a tree. *Quaterly of Applied Mathematics*, **16**, 607-9.

Boorman, S.A., Olivier, D.C. (1973)—Metrics on spaces of finite trees. *Journal of Mathematical Psychology*, **10**, 26-59.

Brossier, G. (1985)—Approximation des dissimilarités par des arbres additifs. *Mathématiques et Sciences Humaines*, **91**, 5-21.

Brossier, G., LeCalvé, G. (1986)—Analyse des dissimilarités sous l'éclairage $\sqrt{D}$, application à la recherche d'arbres additifs optimaux. In D., Diday *et al* (Eds.): *Data Analysis and Informatics, IV*. Amsterdam: North Holland.

Buneman, P. (1971)—The recovery of trees from measures of dissimilarity. In J. Hodson *et al* (Eds.): *Mathematics in the Archaelogical and Historical Sciences*. Edinbhurg. Edinbhurg University Press.

Buneman, P. (1974)—A note on the metric properties of trees. *Journal of Combinatorial Theory*, **17**, 48-50.

Carroll, J.D., Arabie, P (1980)—Multidimensional Scaling. *Annual Review of Psychology*, **31**, 607-49.

Carroll, J.D., Chang, J.J. (1973)—A method for fitting a class of hierarchical tree structure models to dissimilarity data and its application to some body parts data of Miller's. In *Proceedings of the 81th Annual Convention of the American Psychological Association*, **8**, 1097-8.

Carroll, J.D., Clark, L.A., Desarbo, W.S. (1984)—The representation of three-way proximity data by single and multiple tree structure models. *Journal of Classification*, **1**, 25-74.

Carroll, J.D., Pruzansky, S. (1975)—*Fitting of hierarchical tree structure models*. Paper presented at the US-Japan seminar on Multidimensional Scaling. San-Diego.

Cavalli-Sforza, L.L., Edwards, A.W.F. (1967)—Philogenetic analysis models and estimation procedures. *American Journal of Human Genetic*, **19**, 233-57.

Chaiken, S., Dewdney, A.K., Slater, P.D. (1983)—An optimal diagonal tree code. *SIAM Journal of Applied Discrete Mathematics*, **4**, 424-9.

Colonius, H., Schulze, H.H. (1979)—Repråsentation nichnummerischer Ahlichkeitsdaten durch Baumstrukturen. *Psychologische Beitråge*, **21**, 98-111.

Colonius, H., Schulze, H.H. (1981)—Tree structuress for proximity data. *British Journal of Mathematical and Statistical Psychology*, **34**, 167-80.

Coombs, C.H. (1964)—*A Theory of Data*. New York: Wiley.

Corter, J.E., Tversky, A. (1986)—Extended similarity trees. *Psychometrika*, **51**, 429-451.

Coxon, A.P.M. (1982)—*The User's Guide to Multidimensional Scaling*. London: Heinemann.

Culler, M. (1987)—Group actions on R-trees. *Proceedings of the London Mathematical Society*, **55**.

Cunningham, J.P. (1974)—*Finding an optimal tree realization of a proximity matrix*. Paper presented at the Mathematical Psychology Meeting: Ann Arbor.

Cunningham, J.P. (1978)—Free trees and bidirectional trees as representations of psychological distances. *Journal of Mathematical Psychology*, **17**, 165-88.

Day, W.H.E. (1985)—*Analysis of Quartet Dissimilarity Measures between Undirected Phylogenetic Trees*. Research report #CRM-1315. Department of Computer sciences, Montréal (Canada).

De Soete, G. (1983a)—Are non-metric additive tree representations of numerical proximity data meaningful? *Quality & Quantity*, **17**, 475-8.

De Soete, G. (1983b)—A least-square algorithm for fitting additive trees to proximity data. *Psychometrika*, **48**, 621-6.

De Soete, G. (1984)—Additive-tree representations of incomplete dissimilarity data. *Quality & Quantity*, **8**, 387-97.

Dewdney, A.K. (1979)—Diagonal tree codes. *Information & Control*, **40**, 234-9.

Dobson, J. (1974)—Unrooted tree for numerical taxonomy. *Journal of applied Probability*, **11**, 32-42.

Dress, A. (1984)—Trees, tight extensions of metric spaces, and cohomological dimension of certain groups. *Advances in Mathematics*, **53**, 321-402.

Dress, A., Krüger M. (1987)—Parsimonius phylogenetic trees in metric spaces and simulated annealing. *Advances in Applied Mathematics*, **8**, 8-37.

Eck, R.V., Dayhoff, M.O. (1966)—*Atlas of Protein Sequence and Structure*. New York: National Biomedical Research Foundation.

Estabrook, G.F., McMorris, F., Meacham, C. (1985)—Comparison of undirected phylogenetic trees based on subtree of four evolutionary units. *Systematic Zoology*, **22**, 193-200.

Farris, J.S. (1973)—On comparing the shapes of taxonomic trees, *Systematic Zoology*, **2**, 50-4.

Fitch, W.M. (1981)—A non-sequential method for constructing trees and hierarchical classifications. *Journal of Molecular Evolution*, **18**, 30-7.

Furnas, G.W. (1980)—*Objects and their Features: the Metric Representation of Two-Class Data*. Thesis, Standford University.

Gati, I., Tversky, A. (1984)—Weighting common and distinctive features in perceptual and conceptual judgements. *Cognitive Psychology*, **16**, 341-70.

Goldstein, A.G., Chance, J.E., Gilbert, B. (1984)—Facial stereotypes of good guys and bad guys: a replication and extension. *Memory & Cognition*, **10**, 549-52.

Guénoche, A. (1987)—Etude comparative de cinq algorithmes d'approximation des dissimilarités par des arbres à distance additives. Preprint to appear in: *Mathématiques et Sciences Humaines*

Hodson, F.R., Kendall, D.G., Tautu, P. (Eds.) (1971)— *Mathematics in the Archaelogical and Historical Sciences*. Edinbhurg: Edinbhurg University Press.

Hakimi, S.L., Yau, S.S. (1964)—Distance matrix of a graph and its realizibility. *Quaterly of Applied Mathematics*, **22**, 305-17.

Hartigan, J.A. (1967)—Representations of similarity matrices by trees. *Journal of the American Statistical Association*, **62**, 1140-58.

Hartigan, J.A. (1975)—*Clustering Algorithms*. New York: Wiley.

Henley, N.M. (1969)—A psychological study of the semantics of animal terms. *Journal of Verbal Learning and Verbal Behavior*, **8**, 176-84.

Krantz, D.H., Luce, R.D., Suppes, P., Tversky, A. (1971)—*Foundations of measurement*. New York: Academic Press.

Luce, R.D., Galanter, E. (1963)—Psychophysical scaling. In R.D. Luce, R.R. Luce, E. Galanter (Eds.): *Handbook of Mathematical Psychology*. New York: Wiley.

Luong, X. (1983)—*Voisinage Lâche, Score et Famille Scorante*. Multigraph, Besancon (France).

Luong, X. (1988)—*Méthodes en Analyse Arborée: Algorithmes et Applications*. Thesis: Paris V University.

McMorris, F.R. (1987)—Axioms for consensus functions on undirected phylogenetic trees. *Mathematical Biosciences*, **74**, 17-21.

Meacham, C.A. (1981a)—A manual method for character compatibility analysis. *Taxon*, **30**, 591-600.

Meacham, C.A. (1981b)—A probability measure for character compatibility, *Mathematical Biosciences*, **57**, 1-18.

Osherson, D.N. (1987)—New axioms for the contrast model of similarity. *Journal of Mathematical Psychology*, **31**, 93-103.

Patrinos, A.N., Hakimi, S.L. (1972)—The distance matrix of a graph and its tree realization. *Quaterly of Applied Mathematics*, **30**, 255-69.

Phipps, J.B. (1971)—Dendogram Topology. *Systematic Zoology*, **20**, 306-8.

Pruzansky, S., Tversky, A., Carroll, J.D. (1982)—Spatial *versus* tree representations of proximity data. *Psychometrika*, **47**, 3-24.

Roberts, F.S. (1979)—*Measurement Theory*. New York: Addison-Wesley.

Robinson, D.F. (1971)—Comparison of labeled trees of valency three. *Journal of Combinatorial Theory (B Series)*, **11**, 105-119.

Robinson, D.F., Foulds, L.R. (1981)—Comparison of phyologenetic trees. *Mathematical Biosciences*, **53**, 131-47.

Rosch, E. (1973)—On the internal structure of perceptual and semantic categories. In T.E., Moore (Ed.): *Cognitive Development and the Acquisition of Language*. New York: Academic Press.

Rosch, E. (1975)—Cognitive representations of semantic categories. *Journal of Experimental Psychology: General*, **104**, 192-233.

Rosch, E. (1983)—Prototype classification and logical classification: the two systems. In E.K. Scholnik (Ed.): *New Trends in Conceptual Representations*. Hillsdale: Erlbaum.

Roux, M. (1986)—Représentation d'une dissimilarité par un arbre aux arêtes additives. In D. Diday *et al* (Eds.): *Data Analysis and Informatics IV*. Amsterdam: North Holland.

Sattath, S. (1976)—*An equivalence theorem*. Unpublished note, Hebrew University (Israel).

Sattath, S., Tversky, A. (1977)—Additive similarity trees. *Psychometrika*, **42**, 319-45.

Shepard, R.N. (1974)—Representation of structure in similarity data: problems and prospects. *Psychometrika*, **39**, 373-421.

Shepard, R.N. (1980)—Multidimensional scaling, tree-fitting and clustering. *Science*, **210**, 390-8.

Shepard, R.N., Romney, A.K., Nerlove, S.B. (Eds.) (1974)—*Multidimensional Scaling (2 Vol.)*. New York: Academic Press.

Simões-Prereira, J.M.S. (1967)—A note on the tree-realizability of a distance matrix. *Journal of Combinatorial Theory*, **6**, 303-10.

Smolenskii, Y.A. (1963)—A method for linear recording of graphs. *USSR Computional Mathematics & Mathematical Physics*, **2**, 396-7.

Suppes, P., Zinnes, J.L. (1963)—Basic measurement theory. In R.D., Luce, R.R. Bush, E. Galanter (Eds.): *Handbook of Mathematical Psychology*. New York: Wiley.

Turner, J., Kautz, W.H. (1970)—A survey of progress in graph theory in the Soviet Union. *SIAM Review*, **12**, 17.

Tversky, A. (1977)—Features of similarity. *Psychological Review*, **84**, 327-52.

Tversky, A., Gati, I. (1978)—Studies of similarity. In E. Rosch, B.B. Llyod (Eds.): *Cognition and Categorization*. Hillsdale: Erlbaum.

Tversky, A., Gati, I. (1982)—Similarity, Separability and the triangle inequality. *Psychological Review*, **89**, 123-54.

Waterman, M.S., Smith, T.F. (1978)—On the similarity of dendrograms. *Journal of Theoretical Biology*, **73**, 789-800.

Waterman, M.S., Smith, T.F., Singh, M., Beyer, W.A. (1977)—Additive evolutionary trees. *Journal of theoretical Biology*, **64**, 199-213.

Zaretskii, K. (1965)—Constructing a tree on the basis of a set of distance between the hanging vertices (in Russian). *Uspekhi Mat. Nauk.*, **20**, 90-2.

FINDING THE MINIMAL CHANGE IN A GIVEN TREE

Patrick L. Williams[1] and Walter M. Fitch[2]

[1]School of Veterinary Medicine, North Carolina State University,
Raleigh, NC 27606, USA

[2]Department of Biology, University of Southern California,
Los Angeles, CA 90089-1481, USA

It is a common task in biology to determine the genealogy of species, populations, people, or genes and estimate the condition of the ancestral forms. That is often done for molecules such as proteins and nucleic acids. There are many procedures. I address here only the parsimony procedures which ask, " How can I account for the descent of these various sequences from a common ancestor with the fewest number of changes?" The general problem being addressed is as follows. One has a set of s sequences, each sequence being a linear string of letters from some alphabet. In molecular biology the sequences are either proteins, of which there are 20 letters (amino acids), or nucleic acids, of which there are 4 letters (nucleotides). We assume that the sequences have been aligned by some method so that they are all of the same length, t. It is assumed that these s sequences arose from a common ancestral sequence by a branching process that is properly described as a strictly bifurcating tree, that is, as a graph in which there is one and only one path connecting any two nodes on the tree. The tree has s tips (exterior nodes of degree one), one for each of the s sequences and $s-2$ interior nodes of degree three, plus one node of degree two, called the root, that is the ultimate ancestor, the node at which the branching process began. The edges connecting two adjacent nodes are called branches. The task is to discover for any given tree topology, the minimum amount of change, and its nature, on each branch.

We shall use the nucleic acid form of the problem. There are four ribonucleotides, adenine, cytosine, guanine and uracil which are usually represented by the capital letters A, C, G and U (There are also four deoxyribonucleotides, usually represented similarly but with T, thymine, replacing U). The mutational processes by which the sequences may evolve as they descend from node to node, ultimately arriving at a tip are limited. They are (1), substitution of any one letter for another; (2), the deletion of a letter from the string; and (3) the insertion of a letter into the string. (In reality, indels, insertions and deletions, can occur in blocks of length greater than one but that complication is ignored here.) The indel processes are required because they are known to occur, producing long sequences that are identical except for the inserted or deleted elements. It also means that gaps may have to be inserted into sequences in order to align them homologously, that is, so that the $k-$th nucleotide in each of the s sequences all derive directly from the $k-$th nucleotide in the root sequence. The gaps don't really exist in the sequences observed but are place-holders to achieve an homologous alignment. We assume that alignment has already been accomplished by a suitable procedure. [Procedures for aligning two sequences optimally have been developed by Needleman and Wunsch (1970) and Sellers (1974) and the condition for which they are equivalent demonstrated Smith, Waterman and Fitch (1981). Simultaneously aligning s sequences by these methods goes as the $s-$th power and hence heuristics of various sorts are commonly employed.] The presence of gaps introduces a fifth element which will be represented here by N. It is assumed that any sequence element can be substituted by any other element.

The tree relation for the sequences is assumed to specify as well the relationships of the nucleotides in any arbitrary position k. Thus we can and will restrict ourselves to the problem of determining the minimum change at any one position and obtain the minimum result for the entire tree by summation over all t positions.

Parsimony is simply the description of the criterion by which one judges which set of changes (or which tree) is to be preferred in the absence of knowledge as to which changes (or which tree) is historically correct. Parsimony means that the preferred set of changes (or tree) is the one that requires least amount of change, the fewest number of nucleotide substitutions, to account for the descent of the sequences' nucleotides from the ultimate (root) ancestor.

To find the best tree requires one first to determine the least amount of change on a particular tree. It is then only necessary to repeat the determination over all possible topologies to find the best. [The latter can be a formidable task since the number of unrooted trees for s sequences is $\prod(2i-1)$ for $i = 1, \cdots, s-2$. Ways of reducing this large number, such as branch and bounds methods (Foulds et al, 1979a,b), have been proposed, but the general problem is known to be NP complete.]

The parsimony procedure has two forms. The simplest form (uniform weighting) occurs when all substitutions have the same value, namely one. The second form, non-uniform weighting, permits the value associated with substituting A for G, for example, to differ from that of substituting C for G. Uniform weighting is, of course, a special case of non-uniform weighting but is, in fact, much more easily handled by a totally different procedure.

Uniformly Weighted Parsimony.

The uniform weighting procedure, developed by Fitch (1971) and proved by Hartigan (1972) is the only parsimony procedure currently in common use. It is most easily understood as a set-theoretic, two pass procedure that proceeds in the first pass in a systematic way from the tree tips toward the root and in the second pass proceeds from the root back to the tree tips in reverse order.

Consider the tree shown in figure 1, upper, which shows at the tips the nucleotides in some one position for each the s sequences and a topology (phylogeny) proposed for those sequences. Regard the tip sequence assignments as sets. Choose any ancestral node that has two immediately descendent nodes whose sets are specified and assign to the ancestral node the intersunion of the two descendent sets (see figure 2). The intersunion is the intersection of the two descendent sets if the intersection is not empty, otherwise it is the union of the two descendent sets. The process is repeated $s-1$ times, the last time giving the set assignment of the root. The result of this process is shown in the middle tree of figure 1.

The number of times a union was forced in the above procedure (because the intersection was empty; such nodes should be noted), is the length of the tree and is the minimum number of (nucleotide) substitutions required to evolve from any one of the nucleotides in the character set of the root to the specified nucleotides at the tips. If one only wants to know this information, one may quit at this point. But if one wishes to know what substitutions occurred and where on the tree they occured, then one needs to complete the second (reverse) pass.

Before going to that procedure, however, it may be useful to point out that the length of the tree is independent of the location of the root. If one replaces the root node and its two incident edges by a single edge joining directly the two immediate descendants of the root, all interior nodes are then of degree three and the tree is said to be unrooted. It has $2s - 3$ branches and the root could be located on any one of them. For this reason, the formula for the number of rooted trees is the same as the formula given earlier for the number of unrooted trees except that i now goes up to $s - 1$ instead of $s - 2$ [$2i - 1 = 2s - 3$ when $i = s - 1$]. All $2s - 3$ differently rooted trees have the same length and thus only the equivalent unrooted trees need be searched to find the best.

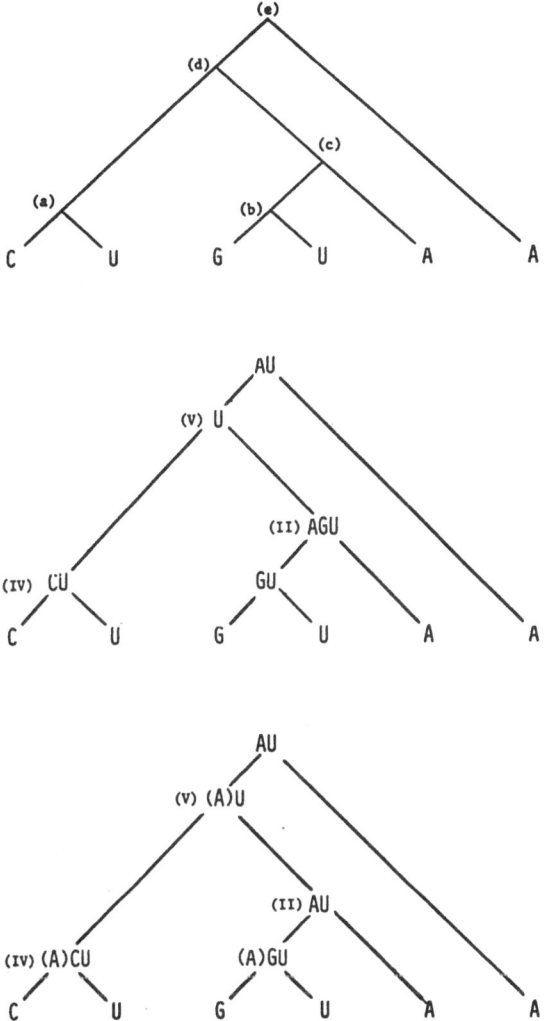

Figure 1: The three steps of uniformly weighted parsimony. The upper tree shows the initial conditons, namely a prescribed tree topology with letters at the tips. The middle tree shows the result after the first pass up the tree using the algorithm in the text. The bottom tree shows the final result after the second pass down the tree by the algorithm in the text. The roman numerals relate to the steps of the second pass algorithm that affect that particular node. Method from Fitch (1971).

COMPARISON OF THE PARSIMONY OPERATION WITH THE UNION AND INTERSECTION OPERATIONS

Intersection $[x,y] \cap [x,z] = [x]$ $[x] \cap [y] = \phi$

Union $[x,y] \cup [x,z] = [x,y,z]$ $[x] \cup [y] = [x,y]$

Parsimony $[x,y] \cup\!\!\!\cap [x,z] = [x]$ $[x] \cup\!\!\!\cap [y] = [x,y]$

The parsimony operation is

   Idempotent $[x] \cup\!\!\!\cap [x] = [x]$

   Commutative $[x] \cup\!\!\!\cap [y] = [y] \cup\!\!\!\cap [x]$

but not

   Associative $\left([x] \cup\!\!\!\cap [x]\right) \cup\!\!\!\cap [y] = [x,y] \neq [x] = [x] \cup\!\!\!\cap \left([x] \cup\!\!\!\cap [y]\right)$

Note that the symbol for the parsimony operation is a combination of the intersection symbol followed by the union symbol, indicating the order of the operations, that is, the union is performed only if the intersection is empty.

Figure 2: The properties of the parsimony operation (intersunion).

By the same token, it is clear that the parsimony criterion cannot indicate the location of the root and, if the determination of the root is important, other criteria must be utilized.

The second pass examines the nodes in the reverse order of the first pass. The set assignments from the first pass are called preliminary. The results of the second pass are final. We take the preliminary root set to be final. The algorithm is as follows.

I. If the preliminary nodal set contains all of the nucleotides present in the final nodal set of its immediate ancestor, go to II, otherwise go to III.

II. Eliminate all nucleotides from the preliminary nodal set that are not present in the final nodal set of its immediate ancestor and go to VI (node c in figure 1, bottom, is an example).

III. If the preliminary nodal set was formed by a union of its descendent sets, go to IV, otherwise go to V.

IV. Add, in brackets, to the preliminary nodal set any nucleotides in the final set of its immediate ancestor that are not present in the preliminary nodal set and go to VI (node a in figure 1, bottom, is an example).

V. Add, in brackets, to the preliminary nodal set any nucleotides not already present provided that they are present in <u>both</u> the final set of the immediate ancestor and in at least one of the two immediately descendent preliminary sets and go to VI (node d in figure 1, bottom is an example).

VI. The preliminary nodal set being examined is now final. Descend one node as long as any preliminary nodal sets remain and return to I above; otherwise quit.

The result of the above procedure is a final set assignment in which every nucleotide present at a node is a possible ancestor in a most parsimonious tree and every nucleotide that is a possible ancestor at a node of a most parsimonious tree is present in that node's set. It does not follow that one can descend from any nucleotide in one final nodal set to any nucleotide in the final nodal set of an immediate descendent. The paths of permissable nucleotide descent are defined as follows.

Brackets denote nucleotides added during steps IV and V above; $i \neq j$.

I. $N_i \rightarrow N_j$ is obligatory if the descendent N_i exists; if the descendent, N_i, does not exist or is bracketed, then and only then-

II. All possible linkages are permitted except $N_i \rightarrow [N_j]$ and $[N_i] \rightarrow [N_j]$.

Rule I asserts that a nucleotide can descend only to itself if its descendant self is not bracketed. Rule two states that, if rule one doesn't apply, the nucleotide can then descend to all nucleotides except bracketed non-selfs.

Non-uniformly Weighted Parsimony.

The basic procedure was first described by Sankoff and Cedergren (1983) and subsequently rediscovered by Williams and Fitch (1988). The procedure of Williams and Fitch is described here. The fundamental question being asked is "At any node, what is the cost for any particular nucleotide to be the ancestor of all the nucleotides at those tips that are descendants of that node?" Here cost is the sum of the substitutions, each times the weight assigned to that particular kind of substitution. The computation of that cost depends solely on having a matrix of substitution values and having the answer to that same question for the two immediately descendent nodes. Thus we can solve the problem for all nodes recursively if we but know the cost at the tip nodes. We make the cost in the tip nodes equal to zero for any nucleotide assigned to that tip and outrageously expensive for any nucleotide not assigned [it is sufficient to make the latter exceed the largest value in the matrix of substitution values].

Since the problem is recursive, it is sufficient to show the method for only three nodes, one of which is the immediate ancestor of the other two. This is shown in figure 3.

The upper node is the immediate ancestor of the lower two nodes, for each of which is shown the cumulative substitution value for each of the four possible ancestral nucleotides. Consider the contribution of the lower left descendent set to the ancestral nucleotide C. That C might best derive from any of the descendent nucleotides. The choice is made by asking that $S(2, j)$, where 2 implies C and j designates the ancestral node, be as small as possible and hence is the minimum of four possible sums $S(i, h) + M(2, i)$ where i goes from 1 to 4, and h designates the (left) descendent node. These four sums are shown on the four lines leading from the four descendent S-cells to the ancestral C cell. The minimum is 24.

The process must be repeated for the four descendent S-cells on the right, for which the minimum is 11. Thus the way to get to $S(2, h)$ with the minimum cost is $24 + 11 = 35$ and hence that value is stored there. The six minima for the other 24 sums are also shown in figure 3.

Each ancestral cell requires the computation of the minimum of two sets of four sums

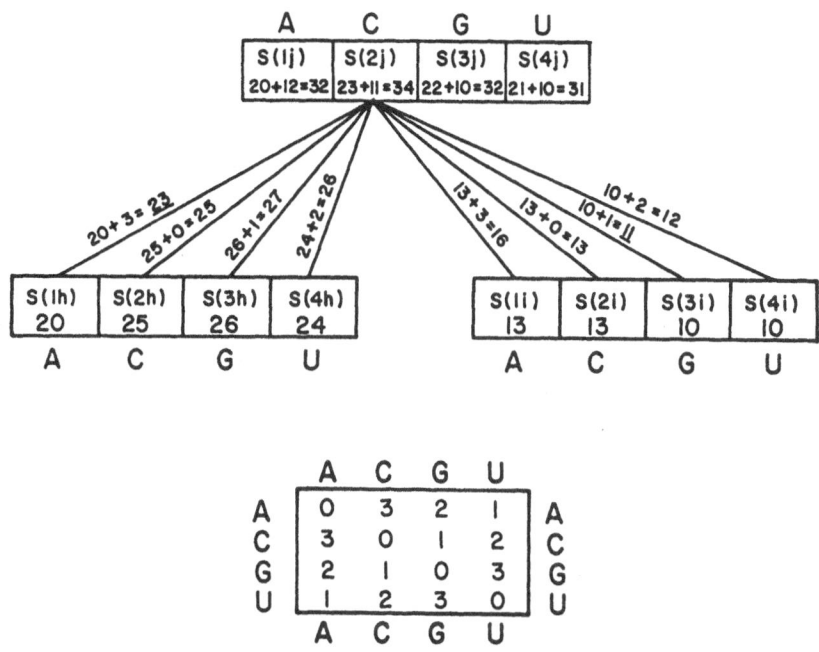

Figure 3: Parsimony procedure when not all changes have the same weight. Consider a region of a tree consisting of an ancestor, j, and its two immediate descendents, h and i and the associated h and i score values, S, for the four letters shown. The problem is to calculate the score values for j using the matrix of substitution values shown at the bottom. The example is shown for the estimate of $S(2, j)$ which is the sum of two minima, each of which is from four sums of two terms representing the score of a descendent nucleotide and the substitution value from C to that descendent nucleotide.

and each ancestor has four cells so that here are eight minima and 32 sums to be determined. For the general case of n different letters possible in the string, there are $2n$ minima and $2n^2$ sums to be determined. For s sequences, this must be repeated for $s-1$ ancestors. If the sequences are t long, then the total computational complexity is $2(s-1)tn^2$ sums of two values to be determined.

The process must be repeated for successively more remote ancestors. After the last ancestor has had its S values determined, the cell(s) with the lowest value represents the score of the tree. If, in figure 3, node h were ancestor $s-1$, then the score of the tree for this position would be 31 and U would be judged the best nucleotide to represent that position. The total score for the tree is obtained by repeating this process over all positions and summing all minimal scores.

The process is clearly generalizable to trees whose ancestors have more than two immediate descendants. It is only necessary to carry the summation over all immediate descendants of the ancestor being examined.

To obtain the subordinate ancestral nucleotides that give rise to the ultimate ancestral nucleotide(s), one could work backwards to see how one got to the U in node h but that is wasteful. It is better, as each ancestor's S set values are determined, to record the nucleotide of each descendant that was used to get the values in the cells of S. In the case of the U in node h of figure 3, it's value was obtained using the A in the left descendent and the U in the right descendent.

Figure 4 shows two different possible trees for the same four sequences as well as two different matrices of substitution values. Each tree is scored for each matrix. The S values for each node are shown, the upper S set deriving from the upper matrix, the lower S set from the lower matrix. The minimum is four for the upper tree only if the upper matrix is used whereas it is four for the lower tree only if the lower matrix is used. Thus among alternative trees, which one is best, which has the lowest score, may depend upon the matrix of substitution values.

Figure 5 shows that the placement of the root can affect the tree score. Thus, unlike the case in simple parsimony, one is not entitled to treat all trees that would be topologically equivalent if the root were removed as having the same score. This behavior is the consequence of the root permitting the insertion between two different nucleotides (here A and U) of a third still different nucleotide (here a C) which proves useful only because the sum of the two substitution values from the third nucleotide is less than the one substitution value between the first two nucleotides. One could make all rooted topology scores independent of the root by making certain that no matrix value, $m_{ij}(i \neq j)$,

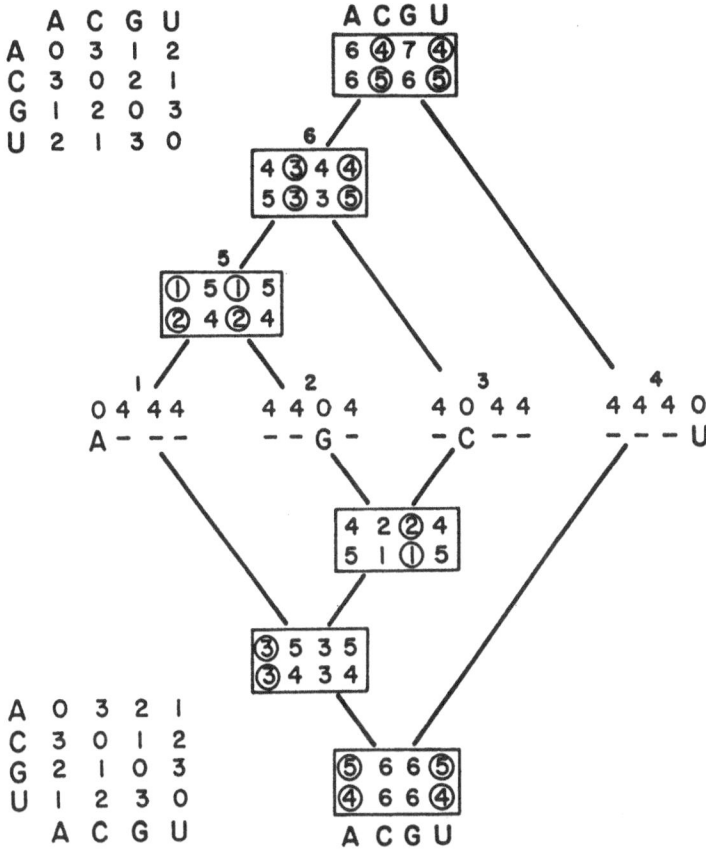

Figure 4: Minimal tree can depend upon the matrix of substitution values. Shown are two trees and, for each, two sets of score values obtained by the procedure in figure 3. The upper set is obtained using the upper left matrix of substitution values, the lower set the lower left matrix. Circled values are for those nucleotides on the path to a minimal score value in the root set. The upper substitution matrix leads to a lower minimal score for the upper tree. The lower matrix gives a lower score for the bottom tree.

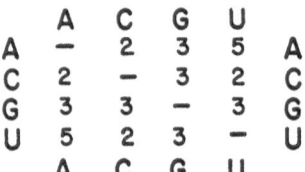

Figure 5: Root location can affect minimal score. Two trees differing only by the location of the root are shown together with a matrix of substitution values used to evaluate the trees. The minimal scores of 4 and 5 (left and right trees respectively) are obtained as shown.

was less than half of any other.

One also would like to know how many times each kind of change occurred in the various parts of the tree. Figure 6 shows that paths of descent for the two trees in figure 4. It shows upper paths originating at both C and U since they have the same minimal score. We take them as having equal expectation and so their frequencies are set at 0.5 each. The lines trace the paths of descent. The unbracketed numbers next to the line are the substitution values. The numbers in brackets are the sum of the frequency of all path's that end on a descendent nucleotide different from the one it started at. That frequency is the same as the frequency of the nucleotide in that branch's ancestor if that nucleotide can descend along only one path to a given descendent. If that nucleotide can descend along more than one path to nucleotides in a given descendent, it divides its frequency equally among all paths.

In figure 6, upper, there is only one branch for which the descent can be along more than one path and that is in the descent from the root U to the penultimate ancestor where the path either terminates on a U or a C. Since the ancestral U frequency is a half, then each of the two paths contributes 0.25 to the total movement from root to penultimate ancestor. There is only one path from the C in the root to the penultimate ancestor and so that path contributes 0.5 to the total which must sum to 1.0 for each branch.

Note the path choices in a branch alters the frequencies of the descendant nucleotides. C and U were 0.5 each in the root but, because of the $0.25U \rightarrow C$ change on the branch to the penultimate ancestor the frequency of C becomes 0.75 while that of U becomes 0.25 in the penultimate ancestor.

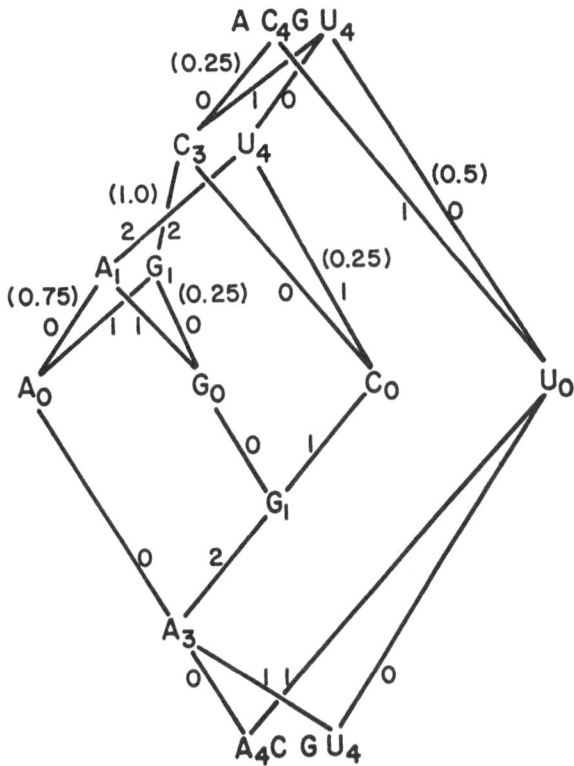

Figure 6: Paths of descent from root nucleotides of minimal score. These are for the minimal trees in figure 4. The four root nucleotides are shown with subscripts for the minimal scores. Lines join nucleotides associated with the minimal score. Unbracketed branches are the substitutions values. Number in parentheses are the fraction of the nucleotides that change to a different nucleotide between ancestor and descendent.

Dynamic Weighting.

The question arises as to what values should be assigned to the various substitutional changes. We pose one solution here (other variants are given in Williams and Fitch, 1988). It presupposes that one is examining many trees to find the best and, having chosen one, is prepared to use it as a basis for changing the matrix of substitution values. We will assume that the original matrix was chosen according to some arbitrary procedure; perhaps it was a uniform matrix with all off-diagonal values equal to one. See Williams and Fitch (1988) for various ways of setting the initial matrix.

On what basis might one wish to differentially weight different substitutions? It is well understood that positions that change frequently are unreliable indicators of relationship and hence one could weight positions inversely to the frequency of their substitutions. While this is easily done, it is not an answer to the question posed. There is however, a corollary to it that asserts that rare changes should be given more weight. Clearly, if nature invented vertebrae only once, then the presence or absence of vertebrae are conclusive in that the possession of the derived character state, vertebrae, by two animals means that they are with certainty more closely related to each other than either is to an invertebrate (this presupposes animals can't lose all their vertebrae under the force of selection). The analog in this case is to examine the frequency with which the various nucleotide changes occurred in the best tree, weight the changes inversely to their frequency of occurrence and then redo the analysis. This we do.

It may be argued that weighting in accordance to the best tree one has just found is biasing the result toward getting the same tree again. However, one frequently observes that a different tree is found so that the bias is not too great to prevent finding what should be a better tree if the premise of weighting inversely to frequency is correct. The process seems always to converge in a few cycles. This procedure has the advantage that it is not biased in its choice of substitutional values other than in setting their initial values.

Discussion.

We are only now trying our program out on various data sets. We present here a few impressions.

One. It seems to be working well in that it elongates the tail of the distribution at the low end suggesting that the weighting is discriminating better between the biological information and the background noise.

Two. It seems to be working well in that it increases the degree of differentness among competing topologies.

Three. It seems to be working well in that it seems to give more often the trees that people are increasingly confident of.

Four. Three is a dangerous preposition in that community belief is a unreliable hook upon which to hang the correctness of a result.

Five. The procedure is sensitive to the tree one gives it to start with.

Six. The procedure is sensitive to the matrix of substitutional values one gives it to start with.

Seven. The procedure is sensitive to the form of the sequences one gives it to start with. It is customary to exclude all positions that have the same nucleotide (or amino acid) in all s sequences. They are obviously uninformative except in maximum likelihood procedures. It is customary in uniformly weighted parsimony to discard as well those positions where all differences involve a unique nucleotide. If all the nucleotides in the k-th position are A except for one G and one U, then this position cannot affect the choice of tree topology. But such a position can affect the values in a dynamically weighted substitution matrix and we have observed that it can greatly affect the result. It is unclear whether it is better to include or exclude positions all of whose differences are singular relative to the majority.

References

[1] Fitch, W. M. (1971). Toward Defining the Course of Evolution: Minimum Change for a Specific Tree Topology, Sys. Zool. 20, 406 - 416.

[2] Foulds, L. R., M. D. Hendy and D. Penny. (1972a). A graph theoretic approach to the development of minimal phylogenetic trees. J. Mol. Evol. 13, 127 - 149.

[3] Foulds, L. R., D. Penny and M. D. Hendy. (1979b) A general approach to proving the minimality of phylogenetic trees illustrated by an example with a set of 23 vertebrates. J. Mol. Evol. 13, 151 - 166.

[4] Hartigan, J. A. (1972). Minimum mutation fits to a given tree. Biometrics 29, 53 - 65.

[5] Needleman, S. B. and C. B. Wunsch. (1970). A general method applicable to the search for similarities in the amino acid sequence of two proteins. J. Mol. Biol. 48, 443 - 453.

[6] Sankoff, D. and R. J. Cedergren. (1983). Simultaneous comparison of three or more sequences related by a tree. In: Time warps, string edits and macromolecules: the theory and practice of sequence comparison, D. Sankoff and J. B. Kruskal eds. Addison-Wesley, London, 253 - 263.

[7] Sellers, P. H. (1974). On the theory and computation of ecolutionary distances. SIAM J. Appl. Math. 26, 787 - 793.

[8] Smith, T. F., M. S. Waterman, and W. M. Fitch. (1981). Comparative Biosequence Metrics, J. Mol. Evol., 18, 38 - 46.

[9] Williams, P. L. and W. M. Fitch. (1988). Weighted parsimony: when not all changes have the same value. Mol. Biol. and Evol. (submitted).

SEARCH, PARALLELISM, COMPARISON, AND EVALUATION:

ALGORITHMS FOR EVOLUTIONARY TREES

David Penny[*] and Pauline Penny[+]
*Botany and Zoology, Massey University, Palmerston North, New Zealand
+Apple and Pear Marketing Board, Bolton St., Wellington, New Zealand

In this lecture we discuss four aspects of the construction and testing of evolutionary trees. These are the large number of possible trees, the potential for parallel computing, the need for an objective comparison of trees, and testing hypotheses about 'unique' events that occurred at remote times in the past. In a short space it is not possible to give a complete review of all four subjects. Instead we will concentrate on our own approaches but give references that allow an introduction to the wider literature. More formal details are given in the relevant publications.

First, we should briefly explain our data. The main data we use are sequences of macromolecules, particularly nucleic acids. These have 4 nucleotides (character states or colours) A, C, G, and U (or T). The sequence for each species (or taxon) is a row of the character state matrix (see table 1). The columns represent equivalent nucleotide positions within the sequences.

A. BRANCH AND BOUND METHODS.

In this section we will use 'parsimony' as the criterion for selecting an optimal tree. This criterion chooses the tree with the smallest number of mutations (the minimal tree, minimal mutation tree, or minimal length tree).

For a given tree, it is easy to both compute optimal sequences for the internal nodes, and the minimal number of mutations for that tree (Fitch, 1971). Still there are $(2n-5)!! = 1 \times 3 \times 5 \times \ldots \times (2n-5)$ different (unrooted, binary) trees with n labelled pendant points (Cavalli-Sforza and Edwards, 1967). Hence if n = 20, there are over 10^{20} possible trees. Even calculating a million trees per second would require several million years to test all trees. Can we do better? Because

Table 1. Twenty informative nucleotide
positions for 10 taxa, the nucleotides
are derived from α-crystallin molecules.

1 Monkey	GAACUCACACAAAAGAAGCC
2 Horse	AGGCUCAAACACAUGCAGGC
3 Kangaroo	AAGUCAUCGAAACUACCUAU
4 Opossum	AAGUCAUCGAAACAACCUAU
5 Rodent	GAGCUCUAACGAAUGCAGCC
6 Rabbit	GAACUCAAACGAAUGCAGCC
7 Dog	AGGCUCUAACGCAUGCAGCC
8 Pig	AGGCUCAAACGCAUGCAGCC
9 Human	GAACUCACACAAAAGAAGAC
10 Cow	AGACUCAAACACAUGCAGCC

this, and several related problems, are known to be NP-complete (Graham and Foulds, 1982; Day and Sankoff, 1987), there is no known efficient algorithm to solve this problem in polynomial time (relative to n).

Nevertheless, some computer programs guarantee to have found all minimal length trees for some data with 19 taxa, even though the methods do not calculate directly more than a thousand trees a second. These are branch and bound methods. They combine the two aspects of a 'branching' (backtracking) algorithm that potentially finds all trees, and a 'bound'. (The bound has most frequently been based on the minimal length (parsimony) criterion which selects the tree of minimal length. Other optimality criteria are possible (for example, likelihood) and some progress has been made in combining them with branch and bound methods.) The search is stopped on a branch when the length of the intermediate tree exceeds the bound, the program then backtracks and starts searching other trees.

Three branch and bound methods have been applied to the problem of finding minimal length trees. The first two methods (Hendy and Penny, 1982) act on the taxa (rows) of the data matrix, the third (Penny and Hendy, 1987) acts on the characters (columns).

A1.Branch and bound on Taxa (Hendy and Penny, 1982) - 'Singles' & 'Pairs'.
The method for searching all trees can be understood by examining the 'decomposition' (the removal of taxa) of any labelled tree. Taxa are the pendant points on an unrooted tree and are labelled 1 to n. With the first method ('Singles') taxa are removed in numerical order, starting with the taxon with the highest number. Removing taxa singly is strait forward (figure 1a). However, neighbouring pairs of taxa can also be removed (neighbouring pairs are taxa without an internal edge between them). Each neighbouring pair is given an 'index' which is the number of the lowest taxon in the pair. The pair with the highest index is removed until there remains only a single pair (when the number of taxa (n) is even) or a triple (when n is odd), see figure 1b. Taxon 1 will always be in the subtree (pair or triple) which is left after other pairs have been removed.

Each backtracking algorithm is a reversal of these procedures. The first method (Singles) starts with three taxa forming one unrooted subtree. Taxa are added (in ascending order) to each internal edge of the subtree until either, the length of the subtree exceeds that of the shortest known tree (the bound) or, all taxa are added and a tree is found that does not exceed the bound. The length of a tree or a subtree is found by applying the Fitch (1971) algorithm to each column of data and summing over all columns. The method works because the length of a particular subtree can never decrease as additional taxa are added. This method is available on both the PAUP and PHYLIP packages (see Fink, 1986), as well as from ourselves.

The second backtracking algorithm (Pairs) starts by forming, in turn, all combinations of pairs of taxa (and if n is odd a single triple containing taxon 1). A distance matrix is used to record the

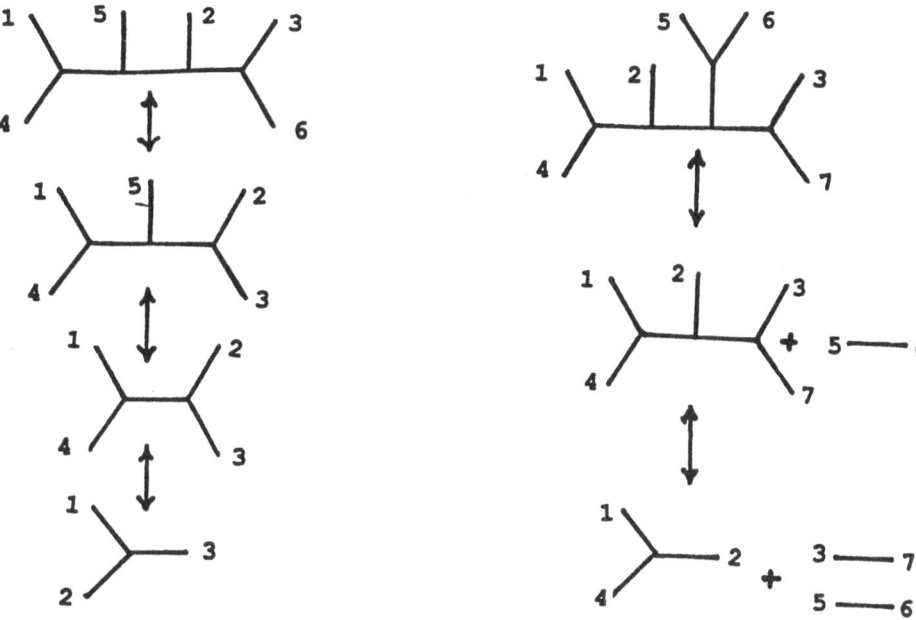

Figure 1. Adding and subtracting singles and pairs of taxa to subtrees.
Examples of a) the 'singles' approach, adding or subtracting taxa singly, and b) the pairs
approach, adding or subtracting taxa in pairs. Each process can be reversed by adding taxa
(singly [a] or in pairs [b]) into each edge of the subtree below it.

distance between each pair of taxa. For a given combination of pairs (and triple) the sum of the
distances is a lower bound on the length of all trees that can be formed from the pairs. If this
value exceeds the length of the shortest known tree it is omitted and another combination of pairs
tested. Otherwise pairs of taxa are joined in turn into the subtree containing taxon one (initially
this will be a pair or a triple).

The length of the subtree is calculated. With sequences this uses the Fitch algorithm (1971) but
an alternative is possible that calculates the length from distance data (Hendy and Penny, in
preparation). With sequences the length of the subtree, together with the distances for pairs not
yet included in the subtree, are added together and checked to see if the sum is greater than the
bound (the shortest tree known). If it is longer than the bound the algorithm backtracks and
another branch is tested. If it does not exceed the bound the procedure is repeated after adding
another pair into an edge of the subtree. This is a reversal of the decomposition algorithm in
figure 1b.

A difficulty with this algorithm is redundancy because most trees can be obtained several ways.
An example is given in figure 2 where there are three ways of finding a tree with six taxa. The
problem becomes worse as n increases so it is essential to eliminate the redundancy. As with the
decomposition algorithm in figure 1b, each pair is given an index which is the lower of the two
taxa in the pair. If the index of the new pair being added is higher than the index of every pair

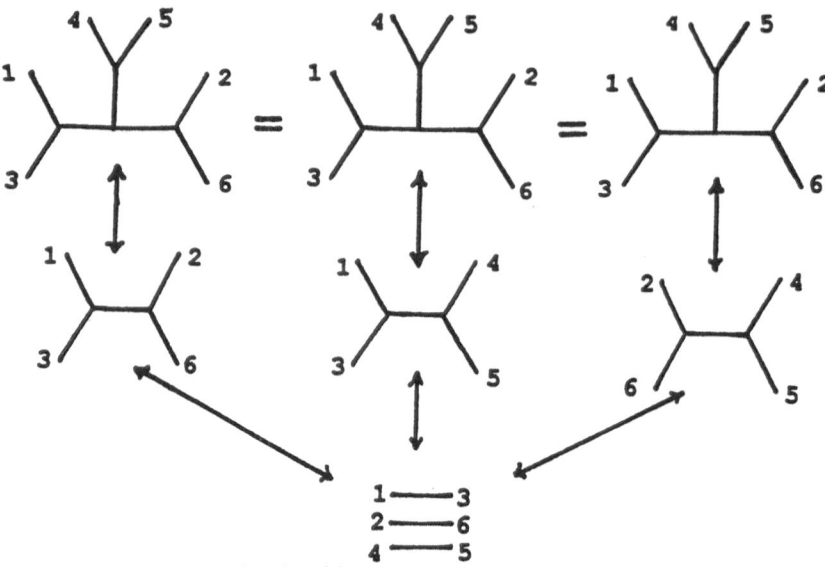

Figure 2. Redundancy in the pairs algorithm.
This one tree with 6 taxa can, in principle, be formed three ways by adding pairs to form larger trees. This redundancy is eliminated by the procedures in Hendy and Penny, (1982).

already in the subtree, then that new pair is added, in turn, to all edges in the subtree. Otherwise, it is added only into the pendant edges of a neighbouring pair of higher index (Hendy and Penny, 1982). Theorems in that paper show that this finds each tree exactly once. This 'pairs' algorithm appeared to have an advantage over 'singles' in that it could get a first estimate of the lower bound using only the distance matrix (which is calculated only once). It was not necessary in the early stages of the algorithm to repeatedly analyse all the columns of the data with each of the possible subtrees.

However, for most data the 'singles' algorithm has proved faster. This appears to result from the many ways of optimizing the 'singles' algorithm. An illustration of its importance is given in Fink (1986) where the same data, the same computer, and the same algorithm (Hendy and Penny, 1982) resulted in running times of 33sec versus 5 hrs! There is still more potential for accelerating the singles algorithm. The main use for the pairs algorithm may be with distance matrices (Hendy and Penny, in preparation) where the same opportunities for optimization do not occur.

A2. Branch and bound on Characters (Penny and Hendy, 1987) - 'TurboTree'.
The previous two methods both work by adding taxa (sequences) into a subtree. A recent development has been a method that includes all taxa from the beginning, but adds columns (nucleotide positions) one at a time. Each column of data, as well as each internal edge of a tree, partitions taxa into disjoint subsets. If two taxa (T_1 and T_2) have nucleotide 'A' at one position, and T_3 and T_4 have 'C', then they can be added into the tree ((1,2)(3,4)) with no 'cost' additional to the one nucleotide change that must occur between 'A' and 'C'. Any other tree breaks the partition (T1,T2), (T3,T4) which is an additional 'cost' to the length of the tree.

The current program uses each column to partition the taxa into subsets, weights the subsets according to the number of times they occur, and then orders the subsets with the highest weights first. These partitions will have the highest cost if they are not included in the tree. The current program has several optimizations for two state characters but additional work is required to optimize it for multistate characters such as nucleotides.

The main point from this section is that we should not allow ourselves to be over-awed by the problem of the numbers of trees. There are data sets where a complete search (such as using branch and bound) is possible and should be used. In many cases it is still not possible to use branch and bound method. In such cases a good 'heuristic' method is useful. These are methods which experience has shown to work well (in that they usually get a solution close to optimal), but they cannot guarantee to have found the optimal solution(s). These methods do introduce another (and unknown) source of error. This additional uncertainty may not be important in some cases but is a serious liability in others. An example is given in section D3 when estimating the reliability of trees by measuring the rate of convergence towards a single tree as longer sequences are used. Here it is essential not to have additional uncertainty introduced by methods that do not find all optimal solutions.

The large numbers of trees is a major problem but we are not at the limit of our abilities for fast searches. A new extension for 'singles' has recently been found (Penny and Hendy, in preparation). It would be particularly helpful to be able to extend the branch and bound approach to other optimisation criteria such as maximum likelihood and related criteria. In addition, a simple relaxation of the branch and bound method gives a reliable and extremely fast heuristic method. There is still another approach that allows identification of an optimal tree without a complete search over all trees (Hendy et al., 1980). This subdivides the columns into subsets for which minimal length trees can be found. The sum of lengths of the subsets is an upper bound on the length of any minimal tree. This method, in one case, has allowed the identification of a minimal length tree with 25 taxa. The method is not generally as useful as the branch and bound methods and has not yet been further developed.

B. PARALLEL COMPUTATION.

The large number of possible trees, and the complexity of calculating likelihood on a single tree, leads to the requirement for massive computing power. This makes the reconstruction of trees a potential subject for parallel processing where many 'computers' undertake the calculations simultaneously.

Approaches to parallel (or concurrent) computing are usually divided in two classes, SIMD (Single Instruction, Multiple Data) and MIMD (Multiple Instruction, Multiple Data). SIMD applies the same instruction simultaneously to many items of data. Coulson et al., (1987) report a very fast SIMD approach for searching a nucleic acid or a protein database for a matching sequence. Their SIMD approach is particularly powerful in that, without assuming random sequences, it finds the probability distribution. Consequently the statistical significance of the results can be more easily interpreted.

In the present case the alternative approach, MIMD, appears more natural. More than one 'processor' is used, each able to run a normal computer program. The processors can be separate computers joined through a network, or a computer with many processing units.

The singles algorithm (section A1) is readily converted to parallel computing in both backtracking and the Fitch algorithm. One aspect of the Fitch algorithm has been implemented in parallel for many years (Fitch and Farris, 1974). This is coding nucleotides as powers of two, A=1, C=2, G=4, U(T)=8. A single step in a program can simultaneously form the union or the intersection of two sets of nucleotides. Here we discuss additional ways of introducing parallelism. Four ways of implementing the singles algorithm in parallel are as follows.

B1) Pipeline. The processors are in a linear array and each carries out one stage of a calculation. Each processor receives results from the previous processor, undertakes one or more steps in the calculation, then passes the result to the following processor. The procedure is repeated with the next item coming down the pipeline. Not all processors are working at the beginning or end of a calculation, but all can be working simultaneously during the middle stages.

With the Fitch (1971) algorithm, the 'preliminary nodal set' is calculated sequentially for each internal point of the tree. Subsequently the 'final nodal set' is calculated for each internal node. A pipeline approach is to have the first processor calculate the preliminary nodal set for the first internal node, passing the result to the next processor which calculates the preliminary nodal set for the second internal node, and so on. Meanwhile, the first processor repeats its calculation with the next column of data.

Table 2. Sequences grouped for multiple processors.

		1	2	3	...	n
1	Monkey	GAACU	CACAC	AAAAG	...	AAGCC
2	Horse	AGGCU	CAAAC	ACAUG	...	CAGGC
3	Kangaroo	AAGUC	AUCGA	AACUA	...	CCUAU
4	Opossum	AAGUC	AUCGA	AACAA	...	CCUAU
5	Rodent	GAGCU	CUAAC	GAAUG	...	CAGCC
6	Rabbit	GAACU	CAAAC	GAAUG	...	CAGCC
7	Dog	AGGCU	CUAAC	GCAUG	...	CAGCC
8	Pig	AGGCU	CAAAC	GCAUG	...	CAGCC
9	Human	GAACU	CACAC	AAAAG	...	AAGAC
10	Cow	AGACU	CAAAC	ACAUG	...	CAGCC

Sequences are shown in groups of five to illustrate parallel computing where there are n processors and the number of columns is not greater than 5n.

B2) The data can be subdivided so that one or more columns (nucleotide positions) are allocated to a processor. An illustration of this is shown in Table 2 with five columns for each processor.The number of columns allocated to each processor need never differ by more than one, ensuring an even balance of work. Communication between processors is reduced because the data (sequences) need only be transmitted once to each processor. Trees are generated in the normal backtracking algorithm, say by the root processor, and are then broadcast to all processors. The lengths for the columns are summed over all processors and checked to see whether the bound is exceeded.

B3) The search over all trees is subdivided between processors. Each processor has all the data and the root processor allocates branches of the tree search to subsidiary processors. Our original program was readily modified for this approach because it allowed checkpointing. That is, at intervals during a run, the information needed to restart the program from that point was stored. The program and data is sent to subsidiary processors that 'restart' part way through the backtracking algorithm. The subsidiary processors only communicate with the root processor when their part of the search is finished, or when a better tree is found.

B4) When minimal length trees are to be found for many sets of data then each set of data can be run on a separate processor. For example, one of the best methods for estimating the reliability of trees is to use repetitive sampling (bootstrapping, jackknifing, and disjoint subsets) (Felsenstein, 1985; Penny and Hendy, 1985; 1986; see section D3). Each sample is run independently on a processor and then a tree comparison metric (section C) is used to compare the results.

In going from alternative B1 to B4 there is a trend from high to low communication between processors. A computer system with a relatively low transmission rate between processors (relative to processor speed) is better suited for the later alternatives. For example, a distributed system (independent computers linked through a local area network) works better with alternatives B3 or B4.

Alternatively, transputers are suitable for any of the alternatives though which alternative is used will be influenced by the amount of local storage available to each transputer. Transputers are described as 'computers on a chip' and are designed to implement the principles of parallelism as defined by Hoare (1985). The most recent models are a single chip containing a central processor, a floating point unit, 4K bytes of fast access memory and four fast communication channels that transmit 10 to 20 Mbytes/sec, independently of the main processor (Anon, 1987).

The pipelining approach described in B1 could be difficult to optimize with transputers because the number required in a pipeline would vary with the number of taxa in the subtree (the number of steps in the Fitch algorithm is approximately equal to twice the number of taxa in the subtree). The number of taxa varies at different stages of the backtracking search, it starts with 4 taxa and ends with n-1. Transputers can be reassigned during processing to avoid them having an unequal load that would leave some of them idle much of the time. However at present reassigning transputers during a run adds considerably to the complexity of a program.

An alternative approach is to have one transputer calculate the preliminary nodal set for each internal point, a second would calculate the final nodal set, and a third would calculate the cost of adding the next taxon to each position in the subtree. An advantage of this approach may be to get both the program and data into the 4K RAM included in each transputer thus eliminating the slower calls to external memory. Larger numbers of transputers could be used in groups of three, essentially combining B1 and B2.

The two options chosen for the first trials were B2 (subdividing the sequences over several transputers) and B3 (subdividing the tree search over several processors). The first has been carried out on a transputer system with an IBM XT as a host. The relevant parts of the program for the branch and bound search have been converted to the OCCAM 2 language (INMOS, 1985) which is the optimal language for transputers. Alternative B3 was implemented by allowing the main program to specify parts of the backtracking search. Subsidiary processors then start from these points and only communicate with the main processor if an improved tree is found.

C. TREE COMPARISON METRICS.

We have found tree comparison metrics allow quantitative tests of evolutionary hypotheses (Penny, et al., 1982; Penny and Hendy, 1985; 1986). To be useful, a tree comparison metric must be easily calculated, should have a known distribution, and preferably be interpretable biologically. Knowing the distribution of the metric helps evaluate if it will be useful for a particular application.

A metric we have found useful, and which meets these three requirements, is the 'symmetric difference' (or partition) metric described by Robinson and Foulds (1981, 1979). Removing an internal edge of a tree partitions the taxa into disjoint subsets. This is repeated for each internal edge and the number of partitions counted that are common (or alternatively, differing) between two trees. This is equivalent to the minimal number of contractions and decontractions (in any order) that must be performed to convert one tree into another (Penny and Hendy, 1985).

This metric is easily calculated (Penny and Hendy, 1985; Day, 1985) which is important because no efficient algorithm is known for some metrics that are potentially useful (for example, nearest neighbour interchange). The method we use to calculate the symmetric difference metric is shown in figure 3. Taxa are coded as powers of two (1,2,4,8...), taxon i becomes 2^{i-1}. The order of the taxa is arbitrary.

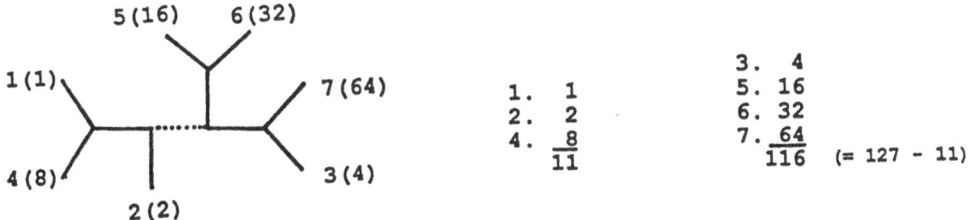

Figure 3. Calculating the symmetric difference tree metric. The tree is shown with one edge removed. This partitions the taxa into disjoint subsets {1,2,4} and {3,5,6,7} which are coded by 11 and 116. The procedure is repeated for each internal edge. The tree can be represented by the numbers 9, 11, 48, and 59.

Removing an internal edge of the tree partitions taxa into subsets and each subset is described by summing the codes of the taxa in the subset. Only the smaller of the two numbers is needed to identify the partition. A binary tree is described uniquely by the n-3 numbers, one for each internal edge. Two trees are compared by counting these numbers that are common (or different). This is not the only method available. Day (1985) has found an elegant algorithm (of order n) to compute this metric.

For binary trees the probability distribution for the symmetric difference metric is known for up to 16 taxa (Hendy et al., 1984; Penny and Hendy, 1986). Some additional results are now available for the behaviour of the metric with very large numbers of taxa (Hendy et al., 1988; Steel, 1988).

The distribution is highly asymmetric with most pairs of trees being the maximal distance apart. This asymmetry can be an advantage, or a disadvantage, depending on the application. It has been an advantage when comparing large numbers of trees from repeated sampling of subsets of columns (see section D3). In this application the trees are very similar and the precision given by the long tail in the distribution is helpful. In other applications the asymmetric distribution could be a disadvantage.

Care is needed when only a small number of trees are compared with this metric because occasionally trees may appear very different, but really only differ in the placing of a single taxon. This can be detected by removing taxa in turn and recomparing the subtrees (Penny and Hendy, 1985). However it is important to develop other tree comparison methods such as quartet measures (Dress, in Eigen and Winkler-Oswatitsch, 1981; Estabrook et al., 1985). These have interesting applications and will become more useful when their distributions are better known. Steel (1989) has recently been able to calculate the mean and variance of the distribution.

D. EVALUATION OF TREE BUILDING METHODS.

This is perhaps the greatest challenge. It has been argued that 'normal' science cannot handle unique events, particularly unique events (such as a divergence) that may have happened hundreds of millions of years ago. We have previously (Penny and Hendy, 1985; 1986) discussed three main approaches to testing the reliability of trees. Here we will expand this to six methods by including one minor method (D6) and splitting what we called the 'empirical' approach into a) probability of fit to a model (for example, likelihood), b) comparing trees from random subsets of columns, and c) congruence, where trees from different sets of data are compared. The main techniques will be discussed from their scope, implementation, strengths and weaknesses.

D1)Simulation.

This is perhaps the most popular method. An ancestral sequence is generated and allowed to 'evolve' by random changes to the sequence. A mechanism of evolution is assumed - this covers aspects such as rates of change on different lineages, whether each site in the sequence has an equal chance of change, and probabilities of interconversion between different nucleotides (or amino acids). Branching (speciation) can be generated randomly, or be predetermined by the program. A set of simulated sequence data is thereby generated which is then used to reconstruct a tree. It is a simple matter to test whether the input tree was recovered. Several authors have used this approach from Peacock and Boulter, 1975 to Li et al., 1987.

There are several limitations with simulation. It is difficult to generalize from a specific simulation (number of taxa, length of sequences, rates of change, restrictions on nucleotide conversion at different sites) to a general case. The mechanism assumed may not accurately reflect evolution in general. The behaviour of a method may change when more taxa are included (some types of errors are reduced) and it does not give the expected behaviour after convergence (when the tree does not change as longer sequences become available). Simulation has its uses but conclusions should be used with caution. We prefer the next three approaches although simulation would be more powerful if combined with these other approaches.

D2)Probability of fit to a model - Likelihood and related methods.

This starts with a model of evolution. In our terminology a model consists of both a tree and a mechanism of change. The mechanism includes the expected rate of change along an edge, the expected distribution of changes along the sequence, and the expected frequency of interconversions between nucleotides (or other data such as amino acids). The model is used to calculate the likelihood that the observed sequences were generated by that model. The model is accepted that gives the highest likelihood of leading to the sequences. Two approaches to its calculation are described by Felsenstein (1982), and Hasegawa et al., (1984)

The use of maximum likelihood approach has developed slowly because, although it has desirable statistical properties, the established methods of calculation are very slow. This is partly because, for even a single tree, each edge on the tree could be assigned many different probabilities of change, in principle, an infinite number for each edge. Obviously, not all possibilities can be tested. The usual method is to use a 'hill-climbing' algorithm that keeps only the edge lengths that lead to an improved fit between observed and expected data. Because of the time involved, most applications of likelihood have not searched more than 10^2 trees whereas branch and bound methods with a parsimony criterion (minimal length) regularly search 10^{15} trees and have searched 10^{19}.

Calculation of maximum likelihood can be accelerated if the molecular clock hypothesis is assumed, this is that the probability of change on each lineage (as well as other aspects of the proposed mechanism) is constant (Lanave et al., 1984; Bishop and Friday, 1985). The clock assumption allows more trees to be calculated because the additional constraints limit the numbers of edge lengths to be calculated. When the assumption is correct then it should help give reliable trees. It is however still necessary to detect cases where rates are genuinely unequal and the molecular clock assumption could then lead to errors.

Accepting the tree with best fit to a model means that the approach is relative, rather than absolute. There may be reasons why a tree is not a good model. For example, if there had been hybridization between taxa then likelihood methods (as well as many others) would still select the tree best supported by the data. Similarly, the mechanism of evolution assumed by the model may be incorrect, again the method will still select a tree that best fits the assumed model. In a Popperian view of science it must always be possible to reject a model. It should be possible, in principle, to say that, given a certain mechanism, no tree will give an acceptable fit to the data.

A new method that promises to overcome some of these difficulties will be described briefly. It does not assume equal rates and it does allow all trees to be rejected for a given mechanism of nucleotide change. This is because it calculates the expected data which can then be compared by standard goodness of fit tests with the observed data. For 2-state characters the method can be summarized by.

$$\underline{q} = -\ln(1-2\underline{p})/2 \qquad (1)$$
$$\underline{r} = K\underline{q} \qquad (2)$$
$$r = \exp(-2\underline{r}) \qquad (3)$$
$$\underline{s} = 2^{1-n}Hr \qquad (4)$$

The vector $\underline{p}$ records the probabilities that two adjacent points on the tree have different codes, $\underline{p}$ has $2n-3$ values. The vector $\underline{s}$ is the frequency of all possible (2^{n-1}) partitions of the n taxa. The partitions are all disjoint subsets of taxa. The Hadamard matrix (H) depends only on the number of taxa and is of size $2^{n-1} \times 2^{n-1}$. K is a matrix of 0's and 1's of columns from H which describes the particular tree being considered.

Matrices H and K have known left inverses so the calculation can be inverted to calculate $\underline{p}$ from $\underline{s}$. The inverse of the Hadamard matrix is unique and involves its transpose (H^t), the inverse is

$H^{-1} = 2^{1-n}.H^t$. K has many possible left inverses and the inverse is chosen that minimizes the sum of squares between the observed and expected values. Full details are in Hendy (1989) and Hendy and Penny (1989).

In the present context it is important that the observed values in $\underline{s}$ (or at an intermediate stage such as $\underline{r}$ or r) can be compared by standard goodness of fit tests. As a consequence it is possible to reject all trees if this does not give an acceptable fit between observed and expected data.

D3)Repetitive sampling (bootstrap, jackknife, and disjoint subsets; rate of convergence). Some newer statistical methods (Efron, 1982) can be applied to the problem of the reliability of the trees. The methods developed for trees (Felsenstein 1985; Penny and Hendy 1985; 1986) form subsets of data by randomly selecting columns, either without replacement (jackknife or disjoint subsets (hobbits)), or with replacement (bootstrap samples). These subsets can have the same number of columns as the original data (bootstrap samples) or fewer columns (jackknife and hobbits). Trees are formed from these subsets by standard methods and the resulting trees are then compared with a tree comparison metric. The two ways of comparing results have been finding internal edges that occur in 95% of the trees (Felsenstein, 1985) or studying the rate of convergence toward a single tree as longer sequences (subsets) are used (Penny and Hendy, 1985;1986).

These approaches allow tentative answers to several interesting questions. Is it likely that another tree will be optimal if more information is gathered for each taxon? Do some methods converge faster than others when longer sequences are used? What trees, other than the optimal, are still reasonable? What size subset of trees is necessary to be confident of including the correct tree?

Is convergence or bootstrapping better? The answer may depend on the application. In a taxonomic study an author may not wish to propose a new taxonomic category unless confident that future work will support it. This is using stability as a criterion. In such a case bootstrapping may be the preferred method since it does not accept an internal edge if there is reasonable doubt. If however the intent is to find the best estimate of the phylogeny (the tree which gives the best prediction as more data becomes available) then convergence appears more suited. It must be noted that although the method allows a decision as to whether convergence has occurred, it is still possible for some methods to converge on an incorrect tree under some circumstances. This is discussed in the next section.

D4)'Analytical' This rather presumptuous title is used in the absence of a better one. The approach attempts to understand the basis of a method and the conditions needed for it to give a correct tree. One simple approach is to calculate the expected data with infinitely long sequences and then test whether a method will recover the original tree. More generally it is possible to evaluate conditions under which a method will lead to an incorrect tree (Cavendar and Felsenstein, 1987; Felsenstein, 1978; Hendy, 1989; Hendy and Penny, 1989).

There are several examples of evolutionary models which generate sequence data where parsimony (minimal length) may converge on the wrong tree, even if infinitely long sequences were available. These include four taxa with unequal rates (Felsenstein, 1978), five taxa with equal rates but unequal lengths (Hendy and Penny, 1989), and six taxa even with all edges being relatively short (Hendy and Penny, 1989). We have summarized this for the parsimony criterion by the maxim "long edges attract". It is still an open question whether the minimal length criterion can be modified (by including parallel changes on adjacent edges) to overcome some of these difficulties. We have found conditions where parsimony will give a correct tree and a simple distance method will give an incorrect tree, and vice versa (Hendy and Penny, 1989).

This 'analytical' approach separates the problem of insufficient data (sampling error) from the question of whether the method will find the correct tree when there is sufficient data to get convergence. This helps understand the strengths and weaknesses of different methods, it is then often possible to modify methods and/or optimality criteria in order to overcome weaknesses.

D5) Congruence - comparison of trees constructed from different data. This is similar to D3 in that trees are constructed with different sequences or sets of data. The difference from D3 is that the sets of data are not randomly selected. The approach has been used for many years with morphological and anatomical characters and has been called congruence (Mickevich, 1978).

An early study with sequences is in Baba et al. (1981) who used seven proteins but made no formal analysis. We compared trees constructed from five different proteins to demonstrate that 'falsifiable' predictions were possible from evolutionary theory (Penny et al., 1982). Other approaches are possible such as building trees separately for the 1st, 2nd, and 3rd nucleotide positions of a coding sequence, from the first half and second half of a sequence, or introns versus exons. Results are then compared to see if the subsets give the same tree.

Testing whether different data give the same tree may be satisfactory with a small number of species where there is a reasonable chance that each set of data will give the same tree. An example is a study of pandas, bears and raccoons (O'Brien et al., 1985).

However, even with 11 taxa we have found that, although trees were highly similar, different sequences did not give the same trees. In one study (Penny and Hendy, 1986) with six proteins, separately and combined, we found 25 minimal trees, all different. To accept any one of these trees as being 'correct' gives at least a 96% chance of both a) rejecting the correct tree, and b) accepting an incorrect one.

Using only the information whether or not two trees were identical would get stuck at this point. But combining sequences, taking random subsets, and using a tree comparison metric (as in D3) allows a much more detailed analysis (Penny, 1985; 1986) than the simple form of congruence used in D5.

D6) Prior knowledge. There are a few cases where evolution has been so rapid that there are records that can be used to test results. Fitch and Atchley (1986) used inbred strains of laboratory mice and in their case each of the methods worked satisfactorily. This example needs to be used with care because Bonhomme et al. (1987) provide additional evidence for a complex origin for the strains of laboratory mice.

Henderson et al., (1987) report a case with influenza viruses. These are RNA viruses and evolve about a million times faster than DNA based organisms. Thus viral isolates from different years change rapidly enough to detect evolution. The results from tree reconstruction were consistent with the knowledge of the times of isolation.

This approach cannot be of general use but, particularly with RNA viruses, may be useful in demonstrating evolution in 'real time', that is on a human time scale.

DISCUSSION

The general problem of the construction and testing of evolutionary trees is both challenging and stimulating. There are challenges for mathematicians, computer scientists, and biologists as well as for those interested in the philosophy of science and/or the limits of science.

There has been a tendency in biologists working with evolutionary trees, perhaps because of the difficulties in testing results quantitatively, to accept lower scientific standards than in physics, chemistry, or in many other areas of biology. This cannot be justified and it is to be hoped that workers studying evolutionary trees will accept only the highest scientific standards.

There is of course still a need to improve existing methods and to develop new ones. New methods can be compared objectively, either by measuring rates of convergence, or by improved understanding of the basis of methods. The potential for accelerating still further branch and bound searches has already been referred to. The application of branch and bound to likelihood calculations is also a top priority.

Parallel computing, in addition to its application to minimal length trees, can be applied to many aspects of the study of trees. There are other calculations where each column of data is handled independently of other columns, examples are with many standard maximal likelihood methods and in calculating distance matrices. With the new likelihood method (Hendy and Penny, 1989) the multiplication by either the Hadamard (H) or K matrices, and by their inverses could be carried out by parallel computers. Both these matrices, and their inverses, become very large as the number of taxa increases, their size is proportional to 2^n.

Several quartet methods have been described that build a tree by examining groups of four taxa (Li, et al., 1987; Penny et al., 1987). The calculation on these quartets, as well as checking whether the results are compatible with a particular tree are also readily handled by parallel processing.

The minimum requirements of a good study may be summarized as follows. For any set of sequences we need to be able to demonstrate objectively that a binary tree is a good representation of the data (as opposed for example to a network - a graph with cycles). It should be possible to suggest a labelled tree that is the best predictor, as well as to indicate the range of trees that could still be possible as more data becomes available. A complete study should include an indication of the mechanism of evolution that generated the sequences, for example, whether all sites were equally likely to change, whether rates of evolution had varied between lineages.

DNA sequencing is becoming more routine and it is even suggested that the work may soon be performed by robots. To make full use of this new data will require a marked increase in our understanding of tree reconstruction methods, together with their strengths and limitations.

Acknowledgements:
We are grateful for assistance, during the preparation of this paper, from members of the University Museum of Zoology, and of the Computer Laboratory, Cambridge University, United Kingdom. We particularly thank Prof. Dress for the invitation to Bielefeld.

References

Anon. 1987. IMS T800 transputer. INMOS Ltd, Bristol

Baba,M.L., L.L.Darga, M.Goodman, and J.Czelusniak. 1981. Evolution of cytochrome c investigated by the maximum parsimony method. J. Molec. Evol. 17: 197-213.

Bishop,M.J. and Friday,A.E. 1985. Evolutionary trees from nucleic acid and protein sequences. Proc. Royal Soc. London,B. 226: 271-302.

Bonhomme,F., J-L.Guenet, B.Dod, K.Moriwaki and G.Bulfield. 1987. The polyphyletic origin of laboratory inbred mice and their rate of evolution. Biol. J. Linn. Soc. 30: 51-58.

Cavalli-Sforza,L.L. and A.W.F.Edwards. 1967. Phylogenetic analysis: models and estimation procedures. Evolution 21: 550-570.

Cavender,J.A. and J.Felsenstein. 1987. Invariants of phylogenies in a simple case with discrete states. J. Classif. 4: 57-71.

Coulson,A.F.W., J.F.Collins, and A.Lyall. 1987. Protein and nucleic acid sequence database searching: a suitable case for parallel processing. Computer J. 30: 420-424.

Day,W.H.E. 1985. Optimal algorithms for comparing trees with labelled leaves. J. Classif. 2: 7-28.

Day,W.H.E. and D.Sankoff. 1987 Computational complexity of inferring phylogenies from chromosome inversion data. J. Theoret. Biol. 124: 213-218.

Efron,B. 1982. The Jackknife, the Bootstrap, and Other Resampling Plans. SIAM monograph No38. Soc. Ind. Appl. Math., Philadelphia.

Eigen,M. and R.Winkler-Oswatitsch. 1981. Transfer-RNA: the early adaptor. Natur-wissenschaften 68: 217-228.

Estabrook,G.F., F.R.McMorris, and C.A.Meachem. 1985. Comparison of undirected phylogenetic tree based on subsets of four evolutionary units. Syst. Zool. 34: 193-200.

Felsenstein,J. 1978. Cases in which parsimony or compatibility methods will be positively misleading. Syst. Zool. 27:401-410.

Felsenstein,J. 1981. Evolutionary trees from DNA sequences: a maximum likelihood approach. J. Molec. Evol. 17:368-376.

Felsenstein,J. 1985. Confidence limits on phylogenies: an approach using the bootstrap. Evolution 39: 783-791.

Fink,W.L. 1986. Microcomputers and phylogenetic analysis. Science 234: 1135-1139.

Fitch,W.M. 1971. Toward defining the course of evolution: minimum change for a specific tree topology. Syst. Zool. 20: 406-416.

Fitch,W.M. and W.R.Atchley. 1985. Evolution in inbred strains of mice appears rapid. Science 228: 1169-1175.

Fitch,W.M. and J.S.Farris. 1974. Evolutionary trees with minimum nucleotide replacements from amino acid sequences. J. Molec. Evolution 3: 263-278.

Graham,R.L., and L.R.Foulds. 1982. Unlikelihood that minimal phylogenies for a realistic biological study can be constructed in reasonable computational time. Mathem. Biosc. 60: 133-142.

Hasegawa,M., H.Kishino, and T.Yano. 1985. Dating of the human-ape splitting by a molecular clock of mitochondrial DNA. J. Molec. Evol. 22:160-174.

Henderson.I.M., D.Penny, and M.D.Hendy. 1987. Models for the origin of influenza viruses. Nature 326:22.

Hendy,M.D. 1989. The relationship between simple evolutionary tree models and observable sequence data. Syst. Zool. (in press)

Hendy,M.D. C.H.C.Little, and D.Penny, 1984. Comparing trees with pendant vertices labelled. S.I.A.M. (Soc.Indust.Appl.Math.) J. Appl. Math. 44: 1054-1067.

Hendy,M.D., L.R.Foulds, and D.Penny, 1980. Proving phylogenetic trees minimal with 1-clustering and set partitioning. Mathem. Biosc. 51: 71-88.

Hendy,M.D. and D.Penny. 1982. Branch and bound algorithms to determine minimal evolutionary trees. Mathem. Biosc. 59: 277-290.

Hendy,M.D. and D.Penny. 1989. A framework for the quantitative study of evolutionary trees. Syst. Zool. (in press)

Hendy,M.D., M.Steel, D.Penny and I.M.Henderson. 1988. Families of trees and consensus. pp 355-362, in (H.H.Bock ed) Classification and Related Methods of Data Analysis, Proc. 1st Conference International Federation Classification Societies (IFCS), Technical University of Aachen. June 29 - July 1, 1987. North-Holland (Amsterdam).

Hoare,C.A.R. 1985. Communicating Sequential Processes. Prentice-Hall, Englewood Cliffs, N.J.

INMOS Ltd. 1984. Occam Programming Manual. Prentice-Hall, Englewood Cliffs, N.J.

Lanave,C., G.Preparata, C.Saccone, and G.Serio. 1984. A new method for calculating evolutionary substitution rates. J. Molec. Evol. 20:86-93.

Li,W.H., K.H.Wolfe, J.Sourdis, and P.M.Sharp. 1987. Reconstruction of phylogenetic trees and estimation of divergence times under nonconstant rates of evolution. Cold Spring Harbor. Symp. Quantit. Biol. 52:847-855.

Mickevich,M. 1978. Taxonomic congruence. Syst. Zool. 27: 143-158.

Peacock,D. and D.Boulter. 1975. Use of amino acid sequence data in phylogeny and evaluation of methods using computer simulation. J. Molec. Biol 95: 513-527.

Penny,D., L.R.Foulds, and M.D.Hendy. 1982. Testing the theory of evolution by comparing phylogenetic trees constructed from 5 different protein sequences. Nature 297: 197-200.

Penny,D. and M.D.Hendy. 1985. The use of tree comparison metrics. Syst. Zool. 34: 75-82.

Penny,D. and M.D.Hendy. 1985. Testing methods of evolutionary tree construction. Cladistics 1: 266-278.

Penny,D. and M.D.Hendy. 1986. Estimating the reliability of evolutionary trees. Molec. Biol. Evol. 3: 403-417.

Penny,D. and M.D.Hendy. 1987. TurboTree: a fast algorithm for minimal length trees. CABIOS (Comp.Applic.BioSci.) 3: 183-188.

Penny,D., M.D.Hendy and I.M.Henderson. 1987. The reliability of evolutionary trees. Cold Spring Harbor. Symp. Quantit. Biol. 52:857-862.

O'Brien,S.J., W.G.Nash, D.E.Wildt, M.E.Bush and R.E.Benveniste. 1985. A molecular solution to the riddle of the giant panda's phylogeny. Nature 317: 140-144.

Robinson.D.F., and L.R.Foulds. 1979. Comparison on weighted labelled trees. Pages 119-126 in Lecture Notes in Mathematics. Vol 748 Springer-Verlag, Berlin.

Robinson.D.F., and L.R.Foulds. 1981. Comparison of phylogenetic trees. Mathem. Biosc. 53: 131-147.

Steel,M.A. 1988. The distribution of the symmetric difference metric on phylogenetic trees. S.I.A.M. J. Discrete Math. 1:541-551.

Steel,M.A. 1989. Distributions on bicoloured evolutionary trees. Ph.D. thesis, Massey University.

THE PHYLOGENY OF PROCHLORON:
IS THERE NUMERICAL EVIDENCE FROM S_{AB} VALUES?
A RESPONSE TO VAN VALEN

Hans-Jürgen Bandelt[1] and Arndt von Haeseler[2]

[1]Fachbereich Mathematik, Carl-von-Ossietzky-Universität,
Postfach 2503, D-2900 Oldenburg, FRG

[2] Forschungsschwerpunkt Mathematisierung, Universität Bielefeld,
Postfach 8640, D-4800 Bielefeld, FRG.

Numerical methods for constructing phylogenies are widely used in molecular biology: often similarity coefficients are derived from (partial) sequence data, and standard clustering algorithms are then applied to produce appropriate dendrograms (Fox *et al.*, 1980; Sneath & Sokal, 1973). Alternative dendrograms based on the same data set may be evaluated by means of numerical criteria. Such criteria do not always seem to be on a sound logical basis: an instance is provided by Van Valen's methodology of evaluating conflicting estimated phylogenies (Van Valen, 1982); the data in question are S_{AB} values for Prochloron, some cyanobacteria and chloroplasts, as reported by Seewaldt & Stackebrandt (1982).

We reinvestigate this data set using a recently proposed criterion (Bandelt & Dress, 1986; Dress *et al.*, 1986; Estabrook *et al.*, 1985) and construct a phylogeny only slightly differing from the average linkage dendrogram (Seewaldt & Stackebrandt, 1982). Nevertheless, in view of the noise present in the data, we do not advocate that this phylogeny (though being optimal in a certain numerical sense) rules out all alternative estimates.

The phylogeny of Prochloron, a symbiont of colonial ascidians, has been a matter of controversial discussion in recent years (Seewaldt & Stackebrandt, 1982; Van Valen, 1982): some biologists favour the idea that Prochloron is a direct descendant of the common ancestor of the chloroplasts and the cyanobacteria, while others regard it as an offshoot of the cyanobacteria (Walsby, 1986).

A comparison of 16S ribosomal RNA of Prochloron, seven cyanobacteria and five chloroplasts reveals that Prochloron is most similar to Fischerella and Nostoc (Seewaldt &

Stackebrandt, 1982). This comparison is based on their S_{AB} values, which are similarity coefficients obtained from partial sequence analysis (Fox *et al.*, 1980). Van Valen (1982), reconsidering these data, claims that the phylogeny he estimates is in better agreement with the data than the average linkage dendrogram derived by Seewaldt & Stackebrandt (1982).

Interestingly, he brings up a numerical argument to justify his conclusion. So, is there really numerical evidence for his point of view?

The ingredients of Van Valen's comparison are the dissimilarity matrix $(D_{ij})(1 \leq i \leq j \leq 13)$ associated with the S_{AB} values (Seewaldt & Stackebrandt, 1982), the ultrametric (U_{ij}) approximating these values in the process of average linkage clustering (*i.e.*, the cophenetic matrix *sensu* Sneath & Sokal, 1973), and a distance matrix (V_{ij}) determined by Van Valen (in a way, however, that is left in the dark); the entries in all three matrices are percent values, ranging from 29 to 84.

There are various ways to measure the goodness of fit to (additive) tree metrics (Bandelt & Dress, 1986); Van Valen has chosen as an objective function the one that sums up the absolute values of the differences between fitted and observed distances. Thus, $\sum \mid U_{ij} - D_{ij} \mid = 157$ is compared to $\sum \mid V_{ij} - D_{ij} \mid = 141$, leading Van Valen to the conclusion that the matrix (V_{ij}) is slightly better than (U_{ij}), whence the same be true for the supporting tree resp. dendrogram.

This is apparently a fallacy since (U_{ij}) is an ultrametric while (V_{ij}) is not. Van Valen intermingles two issues: first, proposing an alternative tree topology, and second, advocating the use of tree metrics instead of just ultrametrics in order to account for the (albeit small) differences in the rate of evolution. The dendrogram of relationship shown in Fig. 1 of the paper by Seewaldt & Stackebrandt (1982) is certainly not intended to represent a best tree fit to their data but to indicate the phylogenetic relationship of Prochloron and its relatives. Of course, the dendrogram can approximate true phylogenetic relationship only in case that the mutational clocks of those organisms are more or less isochronic. This assumption is unavoidable when dealing with clustering methods such as average linkage. One should keep in mind, however, that tree building methods abound that are not affected by different mutation rates. ADDTREE (Sattath & Tversky, 1977) and its modifications (Bandelt & Dress, 1986) constitute pertinent examples, but others may equally well serve the purpose. If one applies the unweighted version of ADDTREE to the matrix (D_{ij}), then yet another tree topology is obtained. Now, the extent to which the three competing tree topologies T_1, T_2, T_3 are supported by the data can be measured in the following way:

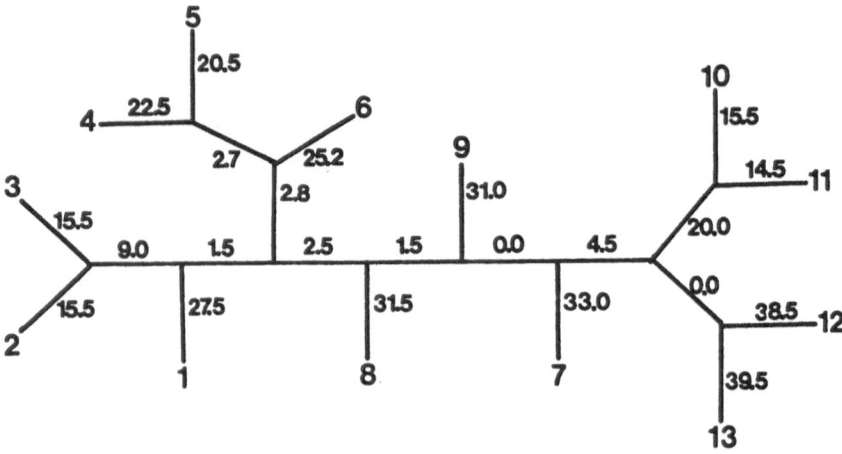

Figure 1: Seewaldt & Stackebrandt's tree T_1. Taxon numbers are: 1, <u>Prochloron</u>; 2, <u>Nostoc</u>; 3, <u>Fischerella</u>; 4, <u>Aphanocapsa 6714</u>; 5, <u>Agmenellum</u>; 6, <u>Aphanocapsa 6701</u>; 7 , <u>Synechococcus 6301</u>; 8, <u>Synechococcus 7502</u>; chloroplasts: 9, <u>Porphyridium</u>; 10, <u>Zea</u>; 11, <u>Lemna</u>; 12, <u>Chlamydomonas</u>; 13, <u>Euglena</u>. The numbers alongside the branches are their estimated lengths according to linear programming.

One assigns lengths to the branches of each topology so that a certain objective function is minimized, and then one compares the three resulting optimal values $\Delta(T_1)$, $\Delta(T_2)$, $\Delta(T_3)$. To this end one uses standard linear programming methods (Waterman *et al.*, 1977). In accord with Van Valen's choice we let the objective function sum up the absolute values of differences. The three trees T_1, T_2, and T_3 are shown in Figs. 1, 2, 3 together with (near-)optimal branch lengths.

Then for the first and third tree topologies we get nearly the same total deviation, namely: $\Delta(T_1) = 122$ and $\Delta(T_3) = 120$, while Van Valen's tree receives $\Delta(T_2) = 140$. We do not want to enter the discussion whether or not the latter difference is well within the limits of error of estimation. But this much can be said:

There is no numerical evidence whatsoever that T_2 is preferable over T_1 or T_3 ! To support this claim one would consider, of course, various other criteria based on different objective functions.

An objective function which may be favoured in this context counts for a proposed tree the number of quartets (groups of four taxa each) whose position in the tree does not

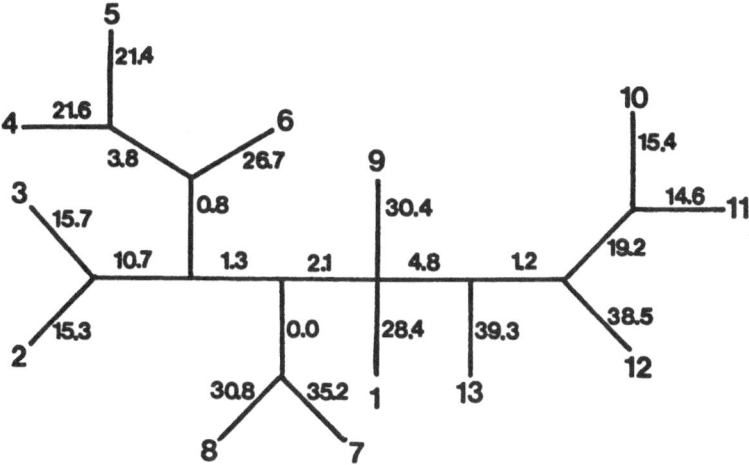

Figure 2: Van Valen's tree T_2 (with slightly modified branch lengths). Taxa are numbered according to Fig. 1

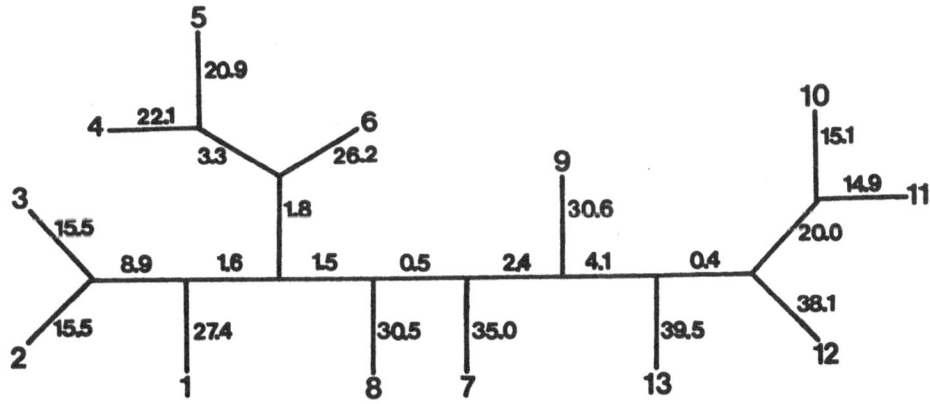

Figure 3: An alternative tree T_3.

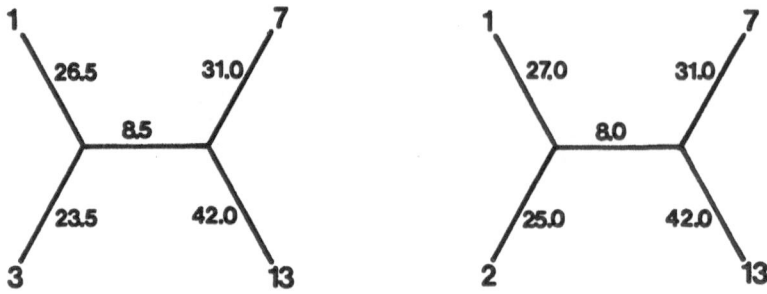

Figure 4: Two quartet trees realizing the observed dissimilarities. Taxa are numbered according to Fig. 1.

reflect the branching pattern supported by the dissimilarities among the four taxa alone. Say, i, j, k, l are four taxa so that the following three dissimilarity sums are ordered as

$$D_{ij} + D_{kl} \le D_{ik} + D_{jl} \le D_{il} + D_{jk}.$$

If the quartet i, j, k, l is in perfect accord with a tree, then, as is well known, the larger two dissimilarity sums are equal. To give an example, consider the taxa 1, 3, 7, 13 or 1, 2, 7, 13 in Fig. 4.

Here the branch lengths add up to the given dissimilarities. Then either quartet tree conflicts with the global tree T_2 (see Fig. 2), but occurs as a subtree within T_1, as well as T_3 - leaving aside branch lengths.

In general, the larger two of the above dissimilarity sums are not necessarily equal, but still as long as $D_{ij} + D_{kl}$ represents the unique smallest sum one would propose the quartet tree where the path (estimated phyletic line) connecting i and j does not cross the one connecting k and l. So, for a global tree tentatively displaying the relationship of all taxa, we count the number of all quartet trees conflicting with the corresponding subtrees derived from the global tree.

Then for the tree T_3 one gets 111 conflicting quartets out of a total of 715 quartets, which amounts to 15.4%. We have tested hundreds of alternative trees via the method of simulated annealing (Dress & Krüger, 1987) in order to minimize the number of conflicting quartets (Dress *et al.*, 1986), so that we are quite confident that T_3 actually represents the unique optimal solution with respect to this criterion. As to the tree T_2, first observe that the 28 quartet trees consisting of the taxa 1, 9 together with one of 10,

11, 12, 13 and one of the remaining taxa each cannot really conflict with T_2 in the above sense because the corresponding subtrees in T_2 are degenerate (*i.e.*, bush-like). Then we find 151 conflicting quartets among 687 quartets, yielding 22.0%. The correponding count for T_1 gives 139 conflicting quartet out of 715 (that is, 19.4%).

Instead of simply counting the overall number of conflicting quartet trees, it is certainly more informative to have a closer look at those quartet trees which consist of Prochloron, one cyanobacterium and two chloroplasts. Then one may also compare the estimated inner branch lengths of such quartet trees. For taxa i, j, k, l with dissimilarity sums being ordered as above, a reasonable estimate for the inner branch length of the supporting quartet tree is given by the following formula

$$\frac{1}{2} \cdot (D_{il} + D_{jk} - D_{ij} - D_{kl})$$

The larger this length is the more confident we are that the individual quartet tree pairing i, j versus k, l (as in Fig. 4 for $i = 1, j = 3, k = 7, l = 13$, say) reflects true phylogenetic relationship, which could be recovered in the global tree. The inner branches in the two trees of Fig. 4 have length 8.5 and 8.0, respectively. Further, in either of the quartet trees with taxa 1, 5, 9, 10 and 1, 5, 9, 11 the taxa 1, 5 are paired versus the other two taxa, the inner branch length being estimated to 5.5. All these instances would certainly suggest to group Prochloron within the cyanobacteria.

On the other hand, taxon 4 (being the closest relative of taxon 5) pairs with taxon 9 versus 1, 10 and 1, 11 respectively. In either case the inner branch length is estimated to 1.5. The remaining four quartet trees (with 1, 10, 8, 9; 1, 11, 8, 9; 1, 12, 8, 9; 1, 12, 7, 13) pairing Prochloron with a chloroplast versus a cyanobacterium and a chloroplast, and thus supporting Van Valen's view, only receive inner branch lengths ranging from 0.5 to 2.0.

Summarizing, we can state that from a numerical point of view the estimated phylogeny of Prochloron including 12 other organisms as proposed by Seewaldt & Stackebrandt (1982) is slightly superior over the one proposed by Van Valen (1982). A tree construction and evaluation method, which is more sophisticated than the usage of average linkage alone, seems to confirm that Prochloron is indeed well within the radiation of cyanobacteria (interstingly, a recently isolated Prochloron-like organism (Burger-Wiersma *et al.*, 1986) containing chlorophylls *a* and *b* is also filamentous - just as Nostoc and Fischerella, which are closest to Prochloron among the cyanobacteria under investigation).

One should, however, be aware of the fact that the present data are not free of considerable noise. This, at least, puts some doubts on the significance of several branches in the tree displayed in Fig. 3, say. A realistic statistical model would provide the appropriate framework for more confident decisions. In any case one is advised to include some more organisms in such an investigation in order to counterbalance the groups of taxa branching off next to the centre of the tree; for example, here it is somewhat unsatisfactory to have only one chloroplast of red algae in the set of taxa.

Finally, the soon available full 16S rRNA and 23S rRNA sequences will shed more light upon the phylogenetic origin of Prochloron. Nevertheless, the principal difficulties in obtaining reliable estimates will pertain. For instance, the distance matrix derived from full 16S rRNA sequences of various archaebacterial species (Woese & Olsen, 1986) contains some internal inconsistencies. Also, the phylogenetic tree which Woese & Olsen (1986) arrive at is not the one with the least number of conflicting quartet trees (the latter number is 10 - from a total of 210 quartets). With full sequences one can estimate quartet trees requiring that the number of necessary nucleotide substitutions be minimized; see Section 8 of Bandelt & Dress, (1986) and cf. Table 3 of Woese & Olsen (1986). Then detecting quartet trees which conflict with an estimated global tree immediately yields a lower bound for the amount of parallelism and reversals that must have occured along the branches; cf. Fitch (1981).

References

[1] Bandelt, H.-J. & Dress, A.: Reconstructing the Shape of a Tree from Observed dissimilarity Data. Advance Appl. Math. 7,309-343 (1986).

[2] Burger-Wiersma, T.; Veenhuis, M.; Korthals, H. J.; Van de Wiel, C. C. M. & Mur, L.R.: A new Prokaryote containing chlorophyll a and b. Nature 320, 262-264 (1986).

[3] Dress, A.; von Haeseler, A. & Krüger, M.: Reconstructing Phylogenetic Trees

using Variants of the Four-Point-Condition. in Studien zur Klassifikation Bd. 17, 299-305 (Indeks Verlag, Frankfurt/Main, 1986).

[4] Dress, A. & Krüger M.: Parsimonious Phylogenetic Trees in Metric Spaces and Simulated Annealing. Advances Appl. Math. 8, 8-37 (1987).

[5] Estabrook, G. F.; McMorris, F.R. & Meacham, C.A.: Comparison of Undirected Phylogenetic Trees based on Subtrees of Four Evolutionary Units. Syst. Zool. 34, 193-200 (1985).

[6] Fitch, W. M.: A Non-Sequential Method for Constructing Trees and Hierarchical Classifications. J. Mol. Evol. 18, 30-37 (1981).

[7] Fox, G.E. et al.: The Phylogeny of Prokaryotes. Science 209, 457-463 (1980).

[8] Sattath, S. & Tversky, A.: Additive Similarity Trees. Psychometrika 42, 319-345 (1977).

[9] Seewaldt, E. & Stackebrandt, E.: Partial sequence of 16S ribosomal RNA and the phylogeny of Prochloron. Nature 295, 618-620 (1982).

[10] Sneath, P. H. & Sokal, R. R.: Numerical taxonomy (Freeman & Co., San Francisco, 1973).

[11] Van Valen, L. M.: Phylogenies in molecular evolution: Prochloron. Nature 298, 493-494 (1982).

[12] Walsby, A. E.: Origins of chloroplasts. Nature 320, 212 (1986).

[13] Waterman, M. S.; Smith, T. F.; Singh, M. & Beyer, W. A.: Additive Evolutionary Trees. J. Theor. Biol. 64, 199-213 (1977).

[14] Woese, C. R. & Olsen, G. J.: Archaebacterial Phylogeny: Perspectives on the Urkingdoms. J. System. Appl. Microbiol. 7, 161-177 (1986).

EVOLUTION OF THE COLLAGEN FIBRIL BY DUPLICATION AND DIVERSIFICATION OF A SMALL PRIMORDIAL EXON UNIT

Hans Hofmann und Klaus Kühn

Max-Planck-Institut für Biochemie

8033 Martinsried bei München

A connection is established between the exon/intron organization of the gene and the aggregation of the molecules of the fiber forming collagen types I, II, and III. The exon structure is found to be reflected in the alternating interaction pattern that governs the D-staggered aggregation of the molecules in the fibril. The origin of the fibril formation is discussed.

1. Introduction

The single collagen molecule (Fig.1, above) consists in its main part of the collagen triple helix which is formed by three peptide chains winding around a common axis. Each chain is about 1000 amino acids long and is composed of tripeptide units of the form Glycine-X-Y, where X and Y stand for arbitrary residues (Fietzek & Kühn, 1976). The repetition of a glycine in every third position is the prerequisite for the formation of a triple helical structure. At both ends the triple helical part bears a short, highly flexible non-helical extension peptide.

In the fibril the molecules aggregate in a staggered array (Hodge & Petruska, 1963), as shown in Figure 1 (below). The stagger distance, named D, is equal to 234 amino acid residues, somewhat less than a quarter of the total molecular length. It is distinguished by a maximum of interaction between the hydrophobic and electrostatic amino acids, as shown by sequence analysis (Hulmes et al., 1973). The lateral arrangement of the molecules in the fibril is still not established. Neighbouring molecules may be staggered with respect to each other by any multiple of D, not only by 1D. At any rate an axially periodic structure is formed. Each repeated layer with a width of D contains five molecular segments, four of full length, named D1 to D4, and the shorter C-terminal 'overlap' segment Ov5. Accordingly, the D-period is divided into an 'overlap' region, containing five

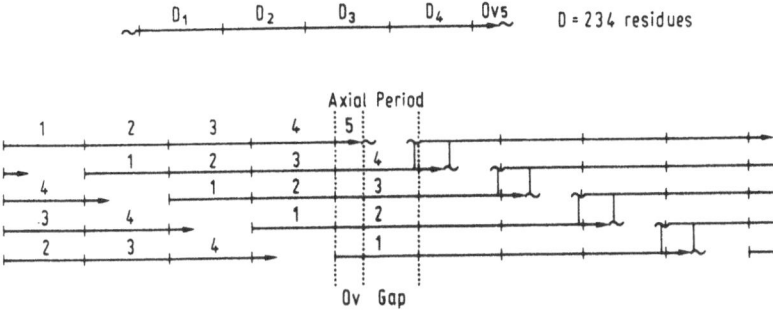

Fig. 1 *Single collagen molecule (above) and Aggregation scheme of the molecules in the fibril (below). The non-helical telopeptides are indicated by wavy lines. For further explanation see text.*

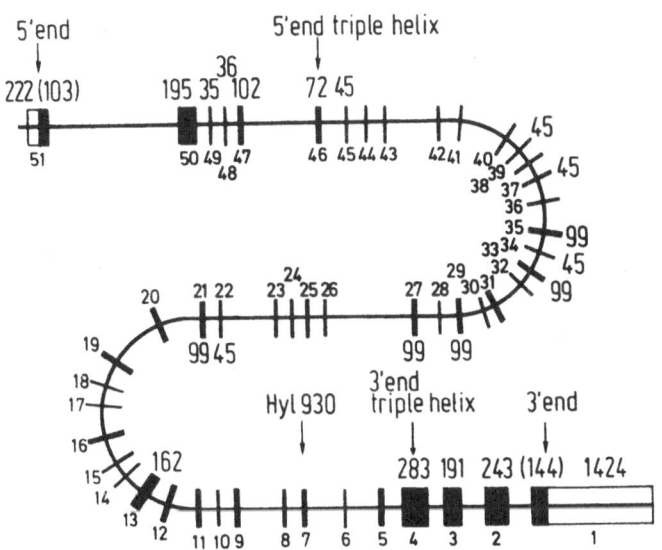

Fig. 2 *Exon/intron organization of the gene of the fiber forming procollagen types I, II, and III. The exons are numbered starting from the 3'end. The length of each exon is given in number of base pairs. In the triple helical part only the shortened exon lengths of 45 and 99 base pairs (see text) are explicitly marked.*

segments, and a 'gap' region with only four. A special head-to-tail bond exists between 4D-staggered molecules which is formed between the non-helical telopeptides and the neighbouring triple helical parts (see Figure 1). This telopeptide/triple helix interaction is based on particularly strong hydrophobic and electrostatic forces, and also a covalent cross link.

The exon/intron organization of the pro-collagen gene is shown in Figure 2 (see e.g. Boedtker et al., 1985). The coding segments, i.e. the exons, are represented by bars. They are embedded into much longer non-coding regions: introns. The globular propeptides at both ends of the procollagen gene are clipped off before fibril formation, with the exception of the short non-helical extension peptides already mentioned. In this work we are dealing with the exons in the triple helical region which show a particular regularity.

In the triple helical domain the exons code for peptide segments of the form

$$(\text{Gly-X-Y})_n, \quad X,Y=\text{arbitrary amino acids.}$$

The occurring lengths n are simply related to each other, as shown in the following table.

Exon length (tripeptides)	Number of exons
6	21
12	9
18	1
5	5
11	11

From this result it was generally concluded that all exon units stem from a common ancestor coding for six tripeptides or 18 residues. Gene fusion yielded exons of double and triple length, and in several cases one tripeptide unit was deleted. It is striking that all the deletions are located in the N-terminal half of the triple helical chain.

Our work deals with the functional significance of the exon/intron organization of the collagen gene, in particular for the D-staggered aggregation of the molecules in the fibril.

2.Results

2.1 In search of a preserved similarity between the exons

We now ask for an inherited similarity that might have been preserved in the exon sequences. A similarity score was calculated for every exon pair, based on Dayhoff's mutatation data (MD) scoring matrix (Dayhoff et al., 1983). In Figure 3 the solid line represents the distribution of scores obtained for the real exons. The broken line marks the average distribution resulting from the random-shuffled exons. The two distributions are essentially indistinguishable, which can also be confirmed by a statistical comparison of their mean values. There are found no high stray scores either that would indicate the sporadic occurrence of extraordinarily similar exon units. Moreover, if all arbitrary sequential stretches of exon size, i.e. with a length of 6 tripeptides, are compared, also no essentially different distribution of similarity scores is produced (not shown).

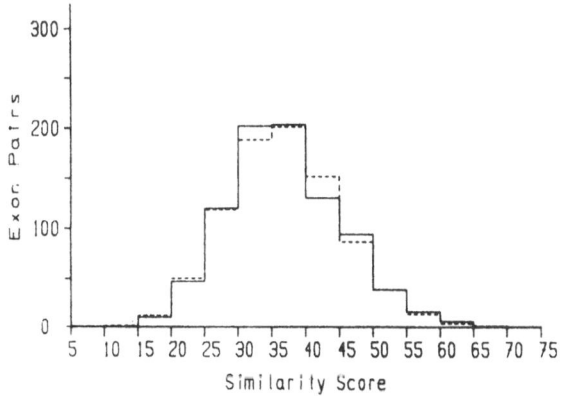

Fig.3 *The distributions of scores obtained for the real and randomly shuffled exon units are compared. M. O. Dayhoff's MD matrix (Dayhoff et al., 1983) was used for the calculation of the scores. All exon units with a length of 18, 36, and 54 residues in the α1 chain of collagen I were taken into account. The units of double and triple length were split into two and three equal parts, respectively. The dashed line represent the mean from 500 random runs.*

Obviously, the similarity between the exons has decreased to a level that is as well attained by arbitrary sequential stretches and random permutations of the exon sequences. An increased average similarity by which the exon set would be distinguished cannot be verified. If the exon structure of the triple helical chain is studied in the framework of the D-staggered aggregation scheme, however, it becomes apparent that original relationships within the exon set must have played a role in the evolution of the latter.

2.2 The exon structure in the framework of the aggregation scheme

Proceeding in this sense, a single axial period of the D-staggered scheme (cf. Fig.1) is shown in Figure 4, and the exon boundaries are marked on the molecular segments D1, ... , D4, and Ov5. One undoubtable connection between the exon frame and the D-staggered association is found in the C-terminal half of the scheme. There is a

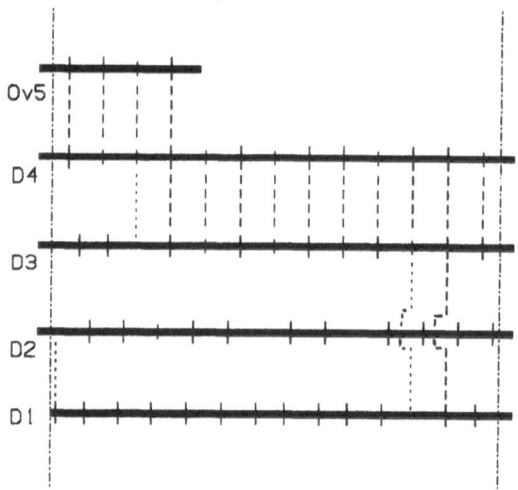

Fig.4 *The exon boundaries are marked (large bars) in the D-staggered aggregation scheme (cf. Fig.1). Small bars divide exon units of double or triple length into equal parts. Coinciding exon boundaries are connected by dashed lines. In certain cases the coincidence depends on the location of the deletion in an exon unit of length 33 residues. Coinciding exon boundaries of this kind are indicated by dotted lines.*

series of exons on the neighbouring molecular segments 3, 4, and 5, which come in phase in the staggered aggregation scheme. (Coinciding exon boundaries are marked by broken lines). We detect a building principle of the fibril which could be called 'the principle of the interacting exon pairs'. It is based on a long sequence of intact exon units in the C-terminal half of the sequence, which is not interrupted by any deletions. This building principle implies the observed fact that the D-stagger is a multiple of the exon size: D = 13e, where e designates the exon size. The regular construction is disturbed, however, by deletions of single tripeptide units, 10 in total, occurring in the N-terminal half of the molecule. Looking at the N-terminal D-segments D1 and D2, only two more local coincidences in the exon frames are detected (see Fig.4).

2.3 The pattern of proline clusters in the aggregation scheme: its relationship to the exon structure

Due to its abundancy (about 1/3) the amino acid proline is distinguished in the triple helical collagen chains. A certain uniformity in the distribution of the proline residues is thought to provide the triple helix with a rather constant stiffness (Hofmann et al., 1984). However, such a structural homogeneity is no longer maintained on the exon scale. Figure 5 shows the fluctuating pattern formed by the proline clusters on the α1 chain of collagen I in the D-staggered scheme. It was produced by moving a representative of a particular proline rich exon, exon 9, along the α-chain and recording similar proline rich stretches. A 'cluster' scoring system was used which counts approximate matches of proline residues (See Legend to Fig.5 and Hofmann and Kühn, 1980, for more detail). In Figure 5 a set of similar exon units of the proline rich type is indicated by peaks that coincide more or less with the corresponding exon boundaries (dashed vertical lines in Fig.5).

Proline clusters tend to occur periodically with the stagger distance of D. Thus proline contact zones are formed between the interacting D-segments D1, ... , D4, and Ov5. In parts of the interaction scheme these zones correspond to 'interacting' exon pairs: The exons 33 to 36 in segment D3 and 41 to 44 in segment D4 in the 'gap' region of the aggregation scheme constitute a pattern of proline rich pairs alternating with proline poor ones. This pattern roughly follows the rythm of the exon frames that coincide between the segments D3 and D4. The exons 15 to 17 in segment D1 extend the same

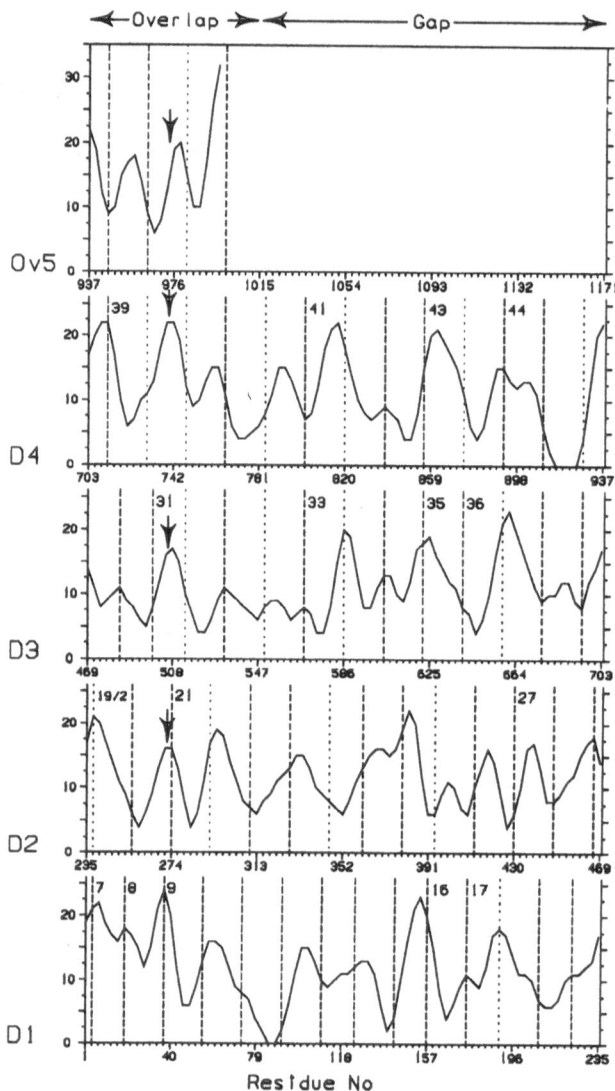

Fig.5 *The distribution of proline clusters on the αl-chain of collagen I in the D-staggered aggregation scheme. Exon 9, as a representative of a particularly proline rich exon unit, was moved along the chain in search of similar proline clusters. The proline 'cluster' score counts a hit if two proline residues on the compared sequential segments differ by not more than 1 tripeptide unit. The dashed vertical lines indicate the exon boundaries. Exons of length 33, 36, and 54 are split into subunits of length 18 and 15 residues, respectively. This is marked by dotted lines. A series of exon numbers is given, starting with exon 7 at the N-terminal end of the triple helical chain.*

pattern in transversal direction. Just in this area the exon boundaries are locally in phase with those in the segments D3 and D4 (cf. Fig.4). On the other hand, the correlation is bad for the proline pattern in the gap region of segment D2, in particular near exon 27, where the exon frame is distinctly displaced with respect to the coinciding exon boundaries in the other D-segments.

Also in the overlap region of the aggregation scheme proline clusters form a specific pattern that reflects parts of the exon structure. The exons 7 and 19/2 (denoting the second half of the long exon 19) as well as 9 and 21/1 are approximately in phase (cf. Fig.4). These exon pairs form proline contact zones between the segments D1 and D2. In contrast, the exon boundaries in the C-terminal half of the overlap region are distinctly displaced with respect to those in the N-terminal half. Nevertheless the proline contact zones formed by the exons in the N-terminal half are extended to layers traversing the whole aggregation scheme (Fig.5). This points to an evolutionary adaptation process in which the proline pattern on the C-terminal segments was assimilated to the pattern preformed by the 'interacting' exon pairs on the N-terminal segments.

2.4 Local periodicities in the proline pattern correspond to exon-exon distances

It is confirmed by prominent repeats with distances of simple multiples of the exon size (e) that an ancestral relationship is still incorporated in the proline pattern. Such repeat distances of 1e, 2e, and 3e have already been detected in former work, when the exon organization was still unknown (Hofmann et al., 1980), and they can now be attributed more precisely to distinct parts of the exon sequence. The proline rich exons 7, 8 and 9 at the N-terminal end of the triple helical chain are connected by a strong 1e repeat (Fig.5). The repeat distance of 2e is associated with the alternating proline pattern described in Figure 5: it connects the exons 16 and 17/2 in segment D1, the exons 19/2 and 21/1 in segments D2, the exons 33/2, 35 and 36/2 in D3, and the exons 41/1, 43/1 and 44 in D4. Spacings of 3e occur between the successive proline clusters formed by the exons 17/2 and 19/2 close to the segment boundary D1/D2, and the exons 36/2 and 39/1 in the neighbourhood of the segment boundary D3/D4 (Fig.5).

A distinct similarity exists also between the 'interacting' exon pairs that constitute the proline contact zones in the D-staggered aggregation scheme (cf. Section 4.3). An inherited similarity in the proline pattern of the exons must have provided a favourable basis for proline-proline interactions between the corresponding triple helical exon segments. It may be due to such an original similarity that the exon frames were brought in phase in the molecular aggregates. The proline-proline interaction might have played a major role in the early aggregates, compared to the electrostatic interaction. The latter requires a complementarity rather than a similarity in the charge patterns, which had but to be acquired in the course of the evolution of the molecular aggregates.

2.5 The alternating interaction pattern

So far nothing was said to characterize the proline poor stretches. To this end, a total view of the interaction pattern was attempted by applying the cluster scoring system to oppositely charged, hydrophobic and proline residues separately. A window of exon size was moved in axial direction between any two of the interacting D-segments D1, ..., D4. Thus a characteristic pattern was found in which proline rich zones alternate with charged layers, roughly according to the rythm of the exon sequence. In a first approximation, the alternating pattern traverses the whole aggregation scheme perpendicular to the fibril axis These results are represented in Figure 6.

The two curves in Figure 6a were obtained by an averaging procedure over all pairs of interacting D-segments D_i and D_j (i,j =1,...,4). The solid line shows the fluctuations in the proline score, as the window is moved along the fibril axis. Interacting charge clusters are indicated by peaks in the broken line. Obviously, the electrostatic interaction tends to evade into the proline poor layers. Hydrophobic contacts, on the other hand, essentially reinforce the proline maxima. This is shown in the Figure 6c, where the sum from the hydrophobic interaction score and the proline cluster score (solid line) is contrasted to the pure proline cluster score (broken line). Only in the lateral neighbourhood of the telopeptides (about residue 79) proline gaps go along with a strong hydrophobicity.

(a)

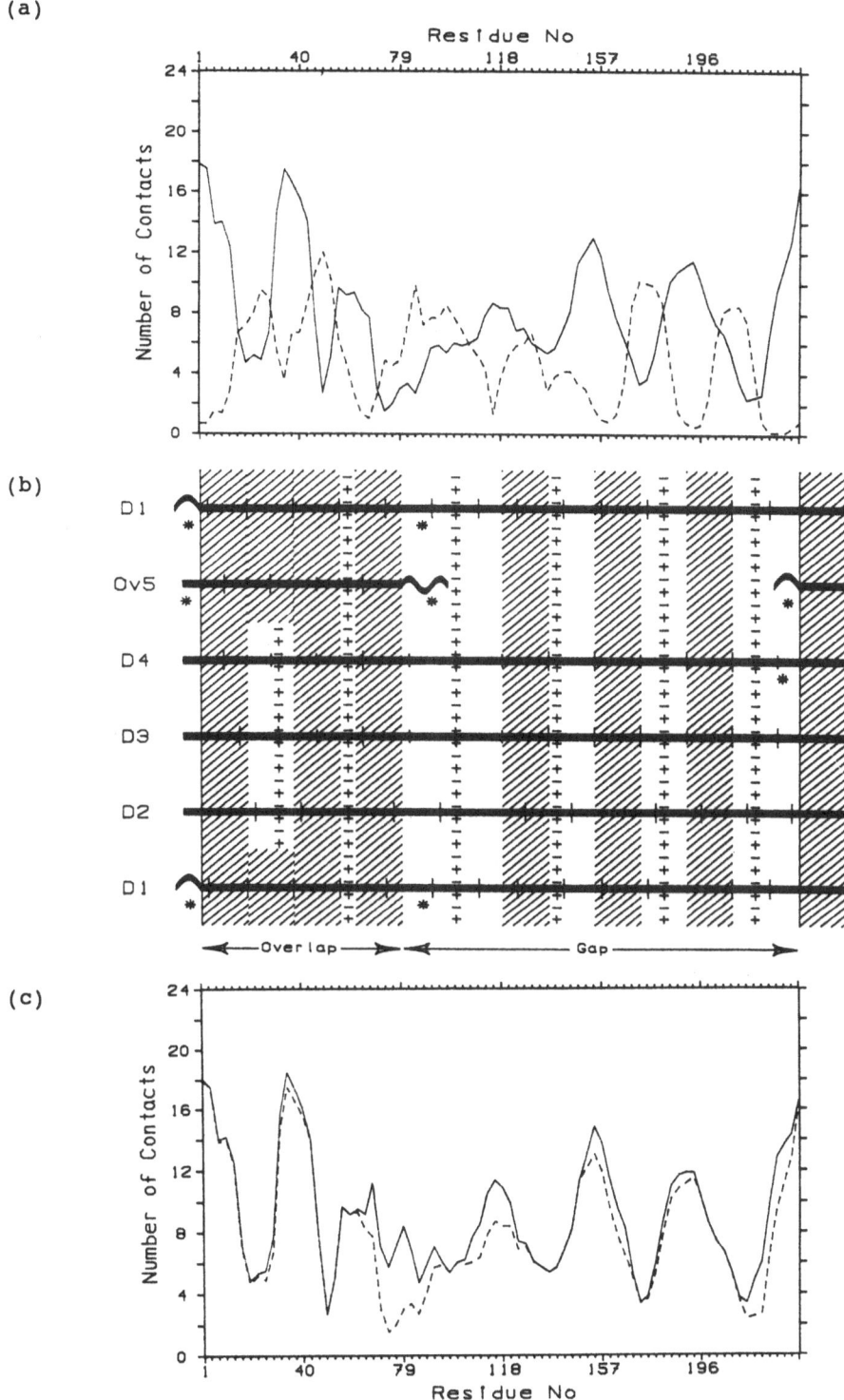

(b)

(c)

Fig.6 (a) *The proline cluster score (solid curve) and the electrostatic interaction score (dashed curve) are shown, as a window of exon width is moved through the aggregation scheme in axial direction. The interaction between all pairs of segments D_i and D_j (i,j =1,...,4) was averaged, but the particularly proline rich end piece Ov5 was excepted. A point of the curve marks the axial position of the left window edge. (b) Schematic representation of the interaction pattern in the D-staggered aggregation scheme. The peaks in the proline curve of Fig.6a coincide with the left edges of the proline rich layers which are represented by cross-hatching. Charged layers are designated by +-. The exon boundaries are marked by short bars in the molecular segments D1,...,D4, and Ov5. The N- and C-terminal telopeptides are depicted by wavy lines, and the asterisks in their neighbourhood mark the cross linking sites. The modified interaction pattern between the 'overlapping' N- and C-terminal end pieces Ov1 and Ov5 is based on separate calculations. (c) The alternating pattern is considerably enhanced by adding the hydrophobic interaction score to the proline cluster score. The solid line corresponds to the combined score. The dashed line represents the pure proline profile, as already shown in (a).*

Figure 6b shows a schematic representation of the alternating interaction pattern, as it can be derived from the curves in the Figures 6a and c. There occur 6 proline rich layers traversing the aggregation scheme. These are cross-hatched in Figure 6b. They are separated by 6 charged layers designated by (+-)-strings. A modified pattern exists between the N- and C-terminal end pieces that constitute the 4D-staggered head-to-tail aggregates. Here one of the charged layers is replaced by a proline contact zone.

There exist certain parts in the aggregation scheme where the interaction pattern is in phase with the exon framework (which is marked by small bars on the molecular segments in Figure 6b). Such regions were already indicated by the proline pattern as analyzed in Section 2.3 (cf. Fig.5). The strongest agreement occurs in the C-terminal half of the gap region, where a series of exon pairs face each other on the neighbouring 1D-staggered segments D3 and D4. Here proline rich exon pairs alternate with charged pairs, which results in a 2e-periodic pattern (e=exon size). At least locally, the exon frame in the gap region of the segment D1 fits also with this pattern, whereas major deviations are found in the segment D2.

In the overlap region the interaction pattern is largely in phase with the exon frames on the segments D1 and D2, but more or less out of phase on the segments D3, D4, and Ov5. The N-terminal half of the proline pattern is formed by a set of similar proline rich exons: exons 7, 8, 9 on the segment D1, and exons 19/2 and 21/1 on the segment D2 (cf. also Fig.5). On the segment D1 the period of the pattern is 1e (e = exon size), and the same is true for 'overlapping' C-terminal end piece Ov5, which thus appears especially assimilated to the N-terminal end piece of the triple helix. On the segment D2 the charge rich exon 20 is inserted between the proline rich exons 19/2 and 21/1 (see also Fig.5). Thus an alternating pattern is preprinted by the exon sequence on D2 that is extended to the segments D3 and D4.

The 2e period (e=exon size) of the alternating pattern is strongly deranged in the overlap/gap interface, including the lateral neighbourhood of the non-helical C-terminal telopeptide (wavy line in Fig.6b). In this region the relationship between the interaction pattern follows not so much the exon framework, but it rather reflects the domain structure of the collagen molecule. Similar features occur also in the gap/overlap interface where the shorter N-terminal telopeptide is located. It is striking that the overlap region is bounded by proline rich layers at both ends. In these the ends of the triple helix are incorporated, in particular the excessivlely proline rich C-terminal tail piece. In contrast, the adjacent regions containing the telopeptides possess a very low proline density. They are rich, in turn, in large hydrophobic and charged residues, occurring in unique residue patterns that are obviously involved in strong telopeptide/triple helix interactions. The latter stabilize the head-to-tail aggregates which play an important role in fibril formation. Obviously, the interaction pattern in the overlap/gap and gap/overlap interfaces is adapted to the requirements of this special type of association. This feature of the interaction pattern is independent of the exon structure and introduces an asymmetry into the periodicity.

3.Discussion

In previous work the origin of the D-staggered aggregation was explained by the repeated duplication of a D-segment (McLachlan, 1976), and the blurred D/6 subperiod of the D-period was considered to serve the lateral association of the molecules and to be imprinted by the helical pitch (Trus and Piez, 1976; Traub, 1978; Hofmann and Kühn, 1980). It can be ruled out that the D-repeat was introduced into the N-terminal half of the triple helical chain by the duplication of entire D-segments, because of the tripeptide deletions located there. However, it is still tempting to postulate a duplication of a whole set of 13 exon units that would have generated two identical D-segments in the C-terminal half of the chain. Between identical segments proline-proline and hydrophobic contacts would be favoured, on the one hand. An identical charge distribution, on the other hand, would rather induce repulsive than attractive electrostatic forces: a circumstance that probably hinders the unstaggered aggregation of the molecules. Actually, the D-staggered aggregation is based on the alternating interaction pattern, as decribed above. This pattern would still remain unexplained by a single duplication event. The few relics of a distinct relationship within the exon set also cannot support a former strict identity of two D-segments. Between the interacting exon pairs in the C-terminal half of the chain a significant similarity is only rarely indicated, and the relationship to other exons is often closer than to the interaction partner. So there is little in favour of a simultaneous duplication of 13 exon units, but it seems also impossible to trace any particular regularity at all in the evolutionary branching process which led to the creation of the triple helical chain.

Definition of the stagger by two interlocking exon sequences in the C-terminal half of the chain

The strongest connection between the exon structure and the staggered association of the molecules is found in a restricted part of the aggregation scheme that comprises the segments D3 and D4 in the gap region of the aggregation scheme. First, the exon frames come in phase in the C-terminal half of two 1D-staggered molecules over the distance of a whole D-segment. But only in the part belonging to the gap region a series of successive proline rich and charge rich exon pairs form an alternating interaction pattern which is also in phase with the coinciding exon frames. Such a pattern appears particularly fit to fix the axial stagger between neighbouring molecules. The close relationship of the interaction pattern to the exon structure suggests that it was not only imprinted by functional constraints, but that it was originally preprinted by the assembly of different exon units. The occurrence of two short interlocking patterns within a longer chain may have been sufficient to provide the corresponding staggered association with a selective advantage. It follows that the stagger distance is a multiple of the exon size: D=13e (e=18 residues).

A genetic origin of the alternating pattern implies that two functionally different building blocks were available for the assembly of the chain: proline rich and charge rich exons. Thus a diversification within the exon set must have preceded the definition of the stagger state. Such a diversification need not have been the result of random modifications only, but a directed functional pressure may already have acted on small precursors of the collagen molecule, even already on single exon units. According to conformational energy calculations by Nemethy and Scheraga (1982), the smallest triple helical entity capable of aggregating in a parallel array is just of exon size. So 'polar pairings' bearing complementary or self-complementary charge patterns may have diverged from the probably original proline rich exon type. The availability of polar building blocks would have ensured that a staggered association was preferred to an unstaggered one.

Possible role of the tripeptide deletions in the N-terminal half of the chain

The relatively short interlocking exon sequences in the C-terminal half of the collagen chain can be regarded as the result of a chance process. Actually, equally favourably interacting exon sequences are missing in the N-terminal half. Apart from local exceptions, the exon frames are out of phase in the corresponding parts of the aggregation scheme. The N-terminal D-segments also form interaction maxima with each other and with the C-terminal D-segments D3 and D4, but here the interaction cannot be founded in the exon structure of the chain. Obviously, the alternating interaction pattern was extended by lateral adaptation of the neighbouring D-segments in large parts of the aggregation scheme. In other parts, showing many deletions, this interaction pattern is distinctly weaker articulated than in the C-terminal half of the gap region. So the interaction between 1D-staggered molecules of collagen I was found to be particularly poor just in the 'gap' subsegment of the molecular segments D1 and D2 (Hofmann et al., 1978). With 7 out of 10 deletions in total, the density of tripeptide deletions is highest here. Altogether, the

N-terminal half of the collagen chain seems to have constituted a more labile association of the molecules which, in consequence, was prone to tripeptide deletions. In contrast, no such deletions were allowed in the C-terminal half, because of its importance for the definition of the D-stagger.

Nevertheless, the exon structure in the N-terminal half still bears a certain relation to the aggregation scheme. Actually, the tripeptide deletions are arranged in a way as to bring at least short parts of the exon frame more or less in phase. This is the case between the 'overlap' subsegments of the molecular segments D1 and D2, where a similarity between proline rich exon units is still incorporated into the interaction pattern. Different from the exon framework in the C-terminal half, the interacting partners are 14 exon units apart instead of 13. The approximate compatibility with the D-stagger is brought about by the 5 deletions in the intervening 'gap' segment. Another local agreement of the exon frames occurs in the 'gap' region. Here the alternating interaction pattern between the segments D3 and D4 is extended to the segment D1, which implies a stabilization of 2D-staggered and 3D-staggered molecular aggregates.

The tripeptide deletions introduce dramatic shifts between the exon frames and so may disturb favourable exon/exon interactions. So one may ask for their benefit in the evolution of the collagen fibril. Has there been a special functional constraint by which he exon frames came out of phase? A possible answer comes from an important special structural element in the aggregation scheme. This is the formation of the 4D-staggered head-to-tail aggregates via the non-helical telopeptides. It is striking that the exon frames are clearly out of phase in the overlapping triple helical end pieces, so that their association cannot be based on relationships between exons. In fact, the formation of the head-to-tail aggregates is mainly determined by strong interactions between the non-helical telopeptides and the neighbouring triple helical parts (Ward et al., 1986). There occur strong hydrophobic and electrostatic forces, and also a covalent crosslink. The unique sequential stretches in the neighbourhood of the telopeptides must have arisen by severe modifications of the original residue pattern. There is no reason why these changes should have respected the exon structure. A shift between the exon frames in the triple helical part of the aggregates may have proved favourable for the establishment of the telopeptide/triple helix interaction. The adaptation of such a shift to the D-periodic scheme implied tripeptide deletions, i.e. a change in the molecular length. Thus the formation of the telopeptide/triple helix interaction meant not only a final stabilization of the already fixed D-staggered aggregation scheme, but a necessary step in the evolution of the aggregation scheme. This corresponds to the behaviour of the collagen molecules in vitro. It was shown by electron microscopical studies that a linear growth step precedes the lateral aggregation in the self-assembly of the collagen molecules, and an orderly fibril formation is not possible without the telopeptides (Ward et al., 1986).

Lateral adaptation of neighbouring molecular segments in the aggregation scheme

After the molecular length was finally fixed, the labile regions in the aggregation scheme suffered no more shifts. So a process of lateral adaptation between the neighbouring molecular segments could take place. In various parts of the aggregation scheme the interaction pattern was extended in transversal direction without respect to the exon structure. Thus proline poor layers were created in the lateral neighbourhood of the telopeptides that are dominated by hydrophobic and electrostatic forces. The distinct phase jump, introduced into the periodicity of the interaction pattern by these layers, distinguishes the D-stagger among other multiples of the exon size.

It was seen that a lateral adaptation must have also been effective in the overlap region of the aggregation scheme. Here proline rich and charge rich layers traverse the whole D-staggered scheme though the exon frames in the N- and C-terminal halves of this region are shifted with respect to each other. Obviously, the interaction pattern on the C-terminal segments was assimilated to the pattern in the N-terminal half. Such lateral adaptations strengthen the 2D- , 3D- , and 4D-staggered association, i.e. the periodicity of the scheme. An evolutionary process appears, at the beginning of which only the 1D-staggered association was supported by the exon structure. The early aggregates were then subjected to the functional constraint to form a periodic 3-dimensional array, which led to the fibril-forming capacity of the collagen molecules.

References

Barsh, G.S., Roush, C.L., Bonadio, J., Byers, P.H. & Gelinas, R.E. (1985). *Proc. Natl. Acad. Sci. USA* **82**, 2870–2874.

Boedtker, H., Finer, M. & Aho, S. (1985). *Ann. N.Y. Acad.* **460**, 85–116.

Dayhoff, M. O., Barker, W.C. & Hunt, L.T. (1983). *Methods Enzymol.* **91**, 524–545.

Fietzek, P.P. & Kuehn, K. (1976). *Intern. Review Conn. Tissue Research* **7**, 1–60.

Hodge, J.A. and Petruska, A.J. (1963). In *Aspects of Protein Structure* (ed. G.N. Ramachandran), Academic Press, New York, 289–300.

Hofmann, H., Fietzek, P.P. & Kuehn, K. (1978). *J. Mol. Biol.* **125**, 137–165.

Hofmann, H. & Kuehn, K. (1980). In *Structural Aspects of Recognition and Assembly in Biological Macromolecules* (eds. M. Balaban, J.L. Sussman, W. Traub & A. Yonath), 427–440.

Hofmann, H., Fietzek, P.P. & Kuehn, K. (1980). *J. Mol. Biol.* **141**, 293–314.

Hofmann, H., Voss, T., Kuehn, K. & Engel, J. (1984). *J. Mol. Biol.* **172**, 325–343.

Hulmes, D.J.S., Miller, A., Parry, D.A.D., Piez, K.A. & Woodhead-Gall- oway, J. (1973). *J. Mol. Biol.* **79**, 137–148.

McLachlan, A.D. (1976). *J. Mol. Biol.* **107**, 159–174.

Nemethy, G. & Scheraga, H.A. (1984). *Biopolymers* **23**, 2781–2799.

Traub, W. (1978). *FEBS L.* **92**, 114–120.

Trus, B.L. & Piez, K.A. (1976). *J. Mol. Biol.* **108**, 705–732.

Ward, N.P., Hulmes, D.J.S. & Chapman, J.A. (1985). *J. Mol. Biol.* **190**, 107–112.

THE POINCARE PARADOX AND THE CLUSTER PROBLEM

Ulrich Höhle

Fachbereich Mathematik, Bergische Universität Wuppertal
Postfach 1000127, D-5600 Wuppertal 1
West Germany

0. Introduction. The motivation behind this paper is to put together two streams of ideas - Poincaré's perception of the physical continuum and the problem to give a characterization of similarity relations in terms of clusters. As the reader will see below both streams encompass the question of finding characteristics of non-transitive systems.

1. Microgeometric spaces.

Opposite to the macrocosm a characteristic of a microgeometric system is the impossibility to attain a complete discernibility of all objects under observation. Therefore in any microgeometric environment a precise identification of individual elements is illusory. K. Menger describes this situation in his famuous paper on "Geometry and Positivism" as follows ([7]): "From the positivistic point of view the difficulty in describing continua does not lie in the use of special kinds of numbers. It lies in the identification of the individual elements,in particular, in their description by numbers of any kind. .. - No one was more keenly aware of the problems connected with the notion of the physical continuum than Poincaré, who emphasized the difficulties on many occasions, e.g. in "La sciences et l'hypothèse", "La valeur de la science" and "Des fondements de la géometrie" . In all these books he characterized the physical continuum by the formulae :

$$A = B \quad , \quad B = C \quad , \quad A \neq C \tag{1.1}$$

symbolizing the fact that an element of a physical continuum may be indistinguishable from two others that can be distinguished from one another. The _mathematical_ continuum, according to Poincaré, is a construct whose main purpose it is to overcome the formulae (1.1) , which he called contradictory or at least repugnant".As a first step towards a mathematically consistent solution of describing continua K. Menger suggests to introduce the concept of lumps as a primitive concept: "I believe that the ultimate solution of the problems of microgeometry lies in a probabilistic theory of hazy lumps" , which "admit an

intermediate stage between indistinguishability and apartness, namely
that of overlapping.....The essential feature of this theory would be
that lumps would not be point sets; nor would they reflect circum-
scribed figures such as ellipsoids. They would rather be in mutual probabilistic relations of overlapping and apartness, from which a metric
would have to be developed" ([7]) .

In the following considerations we present a mathematical model, which
encompasses the formula (1.1) - the so-called Poincaré paradox - and
Menger's idea of overlapping lumbs.

1.1 Definition and remark.

Let T be a continuous t-norm - i.e. a
continuous semigroup operation on the real unit interval subjected to
the following boundary conditions ([10]) :

(i) $T(1,x) = T(x,1) = x$ $\forall\ x \in [0,1]$

(ii) $T(0,x) = T(x,0) = 0$ $\forall\ x \in [0,1]$

A pair (X,E) is called a microgeometric space w.r.t. T iff X is a
set and $E : X \times X \longrightarrow [0,1]$ a map satisfying the following axioms

(MIC1) $T(E(x,x),E(x,y)) = E(x,y)$ (Strictness)
         $T(E(x,y),E(y,y)) = E(x,y)$

(MIC2) $\mathrm{Max}(E(x,x),E(y,y)) \leq E(x,y) \implies x = y$ (Separation)

(MIC3) $E(x,y) = E(y,x)$ (Symmetry)

(MIC4) $T(E(x,y),E(y,z)) \leq E(x,z)$ (Gen. Transitiv.)

1.1.1

We understand X as the carrier of the microgeometric space
(X,E) and associate with E the following interpretation :

1° If $x = y$, then $E(x,x)$ represents the extent to which x exists.

2° If $x \neq y$, then $E(x,y)$ is the extent of overlapping of x and y .

In this context strictness means two things : The extent of existence
is always idempotent w.r.t. the underlying monoidal structure given by
T ; and the extent of overlap cannot exceed the extent to which
elements of the carrier set exist. If the extent of overlapping and
existence coincide, then (MIC2) requires that we should be able to
identify these elements. The meaning of (MIC3) and (MIC4) is obvious.

1.1.2

In the case of Archimedean t-norms the concept of microgeometric
spaces refects the more conservative,positivistic point of view of
microgeometry, namely the idea of the absence of partial existing
elements of the carrier set. Finally in the case of nilpotent t-norms T
every microgeometric space can be provided with an intrinsic relation
$R_E := \{(x,y) \in X \times X \mid 0 < E(x,y)\}$, which is symmetric but not transi-
tive.

1.2 Proposition. Let Δ^+ be the set of all left-continuous, increasing functions $F : \mathbb{R} \longrightarrow [0,1]$ with $F(0) = 0$. Further let T be a continuous t-norm and (X,E) be a microgeometric space (w.r.t. T) provided with the property $E(x,x) = 1$ for all $x \in X$. Then there exists a probabilistic metric $\mathcal{F}_E : X \times X \longrightarrow \Delta^+$ (cf. [10]) satisfying the following conditions

(o) $\qquad T(\mathcal{F}_E(x,y)(r_1) , \mathcal{F}_E(y,z)(r_2)) \leq \mathcal{F}_E(x,y)(r_1+r_2)$

(i) $\qquad \mathcal{F}_E(x,y)(0^+) = E(x,y)$

(ii) $\qquad$ If $x \neq y$, then $\mathcal{F}_E(x,y)(n^+) \leq \sqrt[n+1]{E(x,y)}$ for all $n \in \mathbb{N}$,

$\qquad$ where $\sqrt[n]{\alpha} := \sup \{\lambda , \underbrace{T(\lambda,T(\lambda,\ldots,\lambda))}_{n \text{ times}} \leq \alpha\}$.

Proof. For each pair $(x,y) \in X \times X$ and for each $\alpha \in]0,1[$ let $\mathbb{F}(x,y)(\alpha)$ be the set of all nonnegative integers n determined by the subsequent property

$$\exists (p_k)_{k=0}^{n+1} \in X^{n+2} \text{ with } p_0 = x , p_{n+1} = y , \alpha < E(p_k,p_{k+1}) \ k=0,\ldots,n .$$

Referring to the axioms of a microgeometric space it is not difficult to verify the relations

If $\alpha < E(x,y)$, then $0 \in \mathbb{F}(x,y)(\alpha)$ $\qquad\qquad$ (1.2)

If $n \in \mathbb{F}(x,y)(\alpha)$, then
$\qquad \alpha \leq \sqrt[n+1]{E(x,y)}$. $\qquad\qquad$ (1.3)

$0 \in \mathbb{F}(x,y)(\alpha) \ \forall \ \alpha \in]0,1[\implies x = y$ $\qquad\qquad$ (1.4)

$\mathbb{F}(x,y)(\alpha) = \mathbb{F}(y,x)(\alpha) \ \forall \ (x,y) \in X \times X$ $\qquad\qquad$ (1.5)

If $0 < T(\alpha,\beta)$, then
$\mathbb{F}(x,y)(\alpha) + \mathbb{F}(y,z)(\beta) \subseteq \mathbb{F}(x,z)(T(\alpha,\beta))$ $\qquad\qquad$ (1.6)

Then the left-continuous, increasing function $\mathcal{F}_E(x,y) : \mathbb{R} \longrightarrow [0,1]$ determined by
$[\mathcal{F}_E(x,y)](r) := \sup \{\alpha \in]0,1[\ | \ \mathbb{F}(x,y)(\alpha) \neq \emptyset, \text{ Min } \mathbb{F}(x,y)(\alpha) < r\}$
satisfies the following conditions

$\mathcal{F}_E(x,y)(0) = 0$ $\qquad\qquad$ (1.0')

$E(x,y) = \mathcal{F}_E(x,y)(0^+)$ $\qquad\qquad$ (1.2')

$\mathcal{F}_E(x,y)(n^+) \leq \sqrt[n+1]{E(x,y)}$ $\qquad\qquad$ (1.3')

$\mathcal{F}_E(x,y)(0^+) = 1 \implies x = y$ $\qquad\qquad$ (1.4')

$\mathcal{F}_E(x,y) = \mathcal{F}_E(y,x)$ $\qquad\qquad$ (1.5')

$$T(\,\mathcal{F}_E(x,y)(r_1),\ \mathcal{F}_E(y,z)(r_2)) \quad \leq \quad \mathcal{F}_E(x,z)(r_1+r_2) \tag{1.6'}$$

Hence $\mathcal{F}_E$ is a probabilistic metric on X provided with the desired properties.

By virtue of the preceding proposition the number $E(x,y)$ can also be interpreted as the <u>probability</u> that the distance between x and y is equal to 0 .

2. The cluster problem.

The fundamental problem of cluster analysis consists in characterizing a similarity relation by the set of its clusters in the same way as equivalence relations are uniquely determined by their sets of equivalence classes. The precise situation is as follows : Given a set X of individuals which is not necessarily finite; a measurement process determines a function $S : X \times X \longrightarrow [0,1]$ whose values are interpreted as the degree of similarity between the elements (resp. , individuals) of X . Referring to Jardine and Sibson [6] the similarity is a property of individuals by virtue of their shared attribute states and increases as the number or proportion of shared attribute states increases. Here we assume that a similarity relation fulfills the following axioms

$S(x,x) = 1$	(Reflexivity)
$S(x,y) = S(y,x)$	(Symmetry)
There exists a continuous t-norm T such that for all $x,y,z \in X$ the inequality $$T(S(x,y),S(y,z)) \leq S(x,z)$$ holds.	(Generalized Transitivity)

A "cluster" w.r.t. a given similarity relation S on X is a "subset" of X containing all those elements of X whose similarity is sufficiently large. Two elements are assigned to different clusters if their similarity is sufficiently small (cf. [2]) . In this context the cluster problem consists in an appropriate construction of a set of "clusters" which uniquely determines the underlying similarity relation.

From the mathematical point of view the cluster problem is not well posed : The first question is whether clusters are sets in the usual sense - i.e. in the sense of the classical logic; the second even more delicate question is in which sense clusters are discernible. We only know that the similarity of elements of different clusters is sufficiently small. But what happens to all

those individuals whose similarity is around 1/2 ? In this context
it is remarkable that Jardine and Sibson [6] avoid to give a precise
definition of the notion "cluster" . They explain this situation on
page 40 in [6] as follows : · "Basic to cluster analysis is the
assumption that it is reasonable to seek clusters in the data, these
being subsets of the set" X of individuals "characterized by possession
of the properties of coherence and isolation. Ideal data for cluster
analysis would yield clusters so obvious that they could be picked out
... without the need for complicated mathematical techniques and without making precise what is meant by 'cluster' ".Hence it is natural to
ask these questions : Are clusters and lumps in the sense of K. Menger
the same objects ? Are coherence, isolation, overlapping and apartness
related to each other ?
The theses of this paper are
(i)   Similarity relations are <u>vague</u> equivalence relations.
(ii)   Clusters are <u>vague</u> equivalence classes.
(iii)   The cluster problem is the problem of constructing <u>vague</u>
        quotients w.r.t. vague equivalence relations.

3. The category M-SET.

Now we combine the results of the preceding sections and introduce the
category M-SET (cf. [4]) as follows :   Let   T   be a fixed continuous
t-norm on [0,1] ; then M denotes the completely lattice-ordered monoid
determined by ([0,1],T,$\leq$) . <u>Objects</u> of M-SET are microgeometric
spaces (relative to T) and <u>morphisms</u> of M-SET are structure preserving
maps - i.e. a morphism φ : $(X,E) \longrightarrow (Y,F)$ is an ordinary map
φ : $X \longrightarrow Y$ satisfying the axioms

(M1) $E(x,x)$ = $F(\varphi(x),\varphi(x))$
(M2) $E(x_1,x_2)$ $\leq$ $F(\varphi(x_1),\varphi(x_2))$

The <u>composition</u> is the usual composition of maps and the <u>identity</u> of
(X,E) is the identical map of X .
Anticipating the monoidal structure of M-SET (cf. [4],[5]) we introduce
the concept of weak equivalence relations (resp. , weak equivalence
classes) as follows : Let (X,E) be a microgeometric space, a <u>weak
equivalence relation</u> on (X,E) is a map R : $X \times X \longrightarrow [0,1]$ satisfying
the following axioms :

(R01) $T(E(x,x),R(x,y))$ = $R(x,y)$ (Strictness)
         $T(R(x,y),E(y,y))$ = $R(x,y)$
(R02) $T(R(x,y),E(y,\hat{y}))$ $\leq$ $R(x,\hat{y})$ (Extensionality)
         $T(E(\hat{x},x),R(x,y))$ $\leq$ $R(\hat{x},y)$

(R1)	$E(x,x)$	$\leq$	$R(x,x)$	(Reflexivity)
(R2)	$R(x,y)$	$=$	$R(y,x)$	(Symmetry)
(R3)	$T(R(x,y),R(y,z))$	$\leq$	$R(x,z)$	(Transitivity)

3.1 Remark. (a) Let (X,E) be a microgeometric space. Then E itself is always a weak equivalence relation on (X,E). Because of $E \leq R$ (cf. (MIC1),(RO2),(R1)) E is also called the _discrete_ weak equivalence relation.

(b) If E_c is determined by the Kronecker symbol - i.e. $E_c(x,y) = 1$ iff $x = y$, $E_c(x,y) = 0$ iff $x \neq y$ - then every similarity relation on X is a weak equivalence relation on (X,E_c) .

A _weak equivalence class_ with respect to a weak equivalence relation R on (X,E) is a map $d : X \longrightarrow [0,1]$ provided with the subsequent properties :

(C1)	$T(d(x),\sup_{y \in X} d(y))$	$=$	$d(x)$	(Strictness')
(C2)	$T(d(x),R(x,x))$	$=$	$d(x)$	(Strictness)
(C3)	$T(d(x),R(x,\hat{x}))$	$\leq$	$d(\hat{x})$	(Extensionality)
(C4)	$T(d(x),d(y))$	$\leq$	$R(x,y)$	

In particular each weak equivalence class with respect to the discrete weak equivalence relation E is called a _singleton_ of (X,E) , a concept, which in the case of $T = \text{Min}$ has been introduced by Fourman and Scott in [3] .

3.2 Definition. (Completeness) A microgeometric space (X,E) is complete iff for each singleton s_0 of (X,E) there exists a (unique) element $x_0 \in X$ such that $s_0(y) = E(x_0,y)$ for all $y \in X$.

With regard to the cluster problem we have the following result :

3.3 Theorem. Let CM-SET be the full subcategory of M-SET consisting of all complete microgeometric spaces. If $T = \text{Min}$ or an ordinal sum of nilpotent continuous t-norms (cf. [10]) , then CM-SET has weak quotients - i.e. for every weak equivalence relation in CM-SET the weak quotient exists.

The _proof_ can be found in [4] and [5] ; here we restrict ourselves to the description of the weak quotient and the weak quotient map : Let R be a weak equivalence relation on (X,E) and $\mathbb{E}_R(X,E)$ be the set of all weak equivalence classes w.r.t. R . On $\mathbb{E}_R(X,E)$ we consider an

"equality relation" $\tilde{R}$ defined by $\tilde{R}(d_1,d_2) := \sup_{x \in X} T(d_1(x),d_2(x))$.
Then $(\mathbb{E}_R(X,E),\tilde{R})$ is a complete microgeometric space and
$\varphi : X \longrightarrow \mathbb{E}_R(X,E)$ defined by $[\varphi(x)](y) := R(x,y)$ is an epimorphism.
It follows from [4] and [5] that φ is the weak quotient map and
$(\mathbb{E}_R(X,E),\tilde{R})$ the weak quotient of $(X;E)$ w.r.t. R .

3.4. Remark. (a) In the case of T = Min Theorem 3.3 is well known :
In this case CM-SET is obviously isomorphic to the category of sheaves
over the complete Heyting algebra $[0,1]$; hence CM-SET is a topos
(cf. [3]) . Since in a topos the concepts of equivalence relations in
the usual (i.e. categorical) sense and weak equivalence relations as
well as quotients and weak quotients coincide (cf. [5]) , the theorem
follows from the exactness properties of toposes (cf. [1]) .
(b) Let T be a nilpotent, continuous t-norm and (f,g) be an additive
generator of T (cf. 5.5.2,5.5.4 in [10]) . Then the support set of the
weak quotient w.r.t. a weak equivalence relation R on (X,E) is isomorphic to the completion of X w.r.t. the uniformity generated by the
quasi-metric $g \circ R$.

4. Consequences for cluster analysis

The equivalence between dendograms and quasi-ultra-metrics is one of
the cornerstones of cluster analysis. Unfortunately this equivalence
breaks down in the case of general quasi-metrics. The main reason for
this situation lies in the non-transitivity of the relations

$$r_\alpha := \{(x,y) \in X \times X \mid \rho(x,y) \le \alpha\} \quad , \quad 0 < \alpha \quad .$$

Consequently there exists an extensive literature devoted to the pro-
blem to overcome this difficulty. In this context the solution given in
Theorem 3.3. (see also Remark 3.4) is surprisingly rather simple. But
Theorem 3.3 has even deeper consequences :

(i) In the case of quasi-ultra-metrics (i.e. T = Min) the quotient
has nothing to do with dendograms. As an illustration let us consider
the following
Example. $X = \{0,1\}$, $S(x,y) = \left\{ \begin{array}{l} 1 \; , \; x = y \\ \varkappa \; , \; x \ne y \end{array} \right\}, 0 < \varkappa < 1$. Then ⌐⌐ is
the corresponding dendogram, but the carrier of the quotient is deter-
mined by the set
$$\{(\alpha,\varkappa) \mid \alpha \in [\varkappa,1]\} \cup \{(\varkappa,\alpha) \mid \alpha \in [\varkappa,1]\} \cup \{(\alpha,\alpha) \mid \alpha \in [0,\varkappa]\} \quad .$$

(ii) Clusters are models of sets with respect to certain non-
classical logics - i.e. are maps with codomain $[0,1]$ satisfying w.r.t.
T a list of certain logical axioms ; e.g. in the case of T = Min
clusters are models of sets w.r.t. the intuitionistic logic , while in

the case of nilpotent,continuous t-norms clusters are models of sets w.r.t. the $\aleph_{\tau}$-Łukasiewicz logic.

(iii) The "equality relation" $\tilde{R}$ on $\mathbb{E}_R(X,E)$ can be understood as a measure of overlapping and solves the question , in which sense clusters are discernible.

5. Concluding remark.

Non-transitivity is the connecting link between Poincaré's paradox and the cluster problem. An appropriate, mathematical formulation of this situation is the structure of a microgeometric space, in which Poincaré's paradox is valid as well as the cluster problem has a satis-factory solution. In this context dendograms have only a limited meaning ; it seems to be more reasonable to base the under-standing of vague data on the intrinsic structures of microgeometric spaces - e.g. on probabilistic metrics as indicated in proposition 1.2.

References.

[1] M. Barr and C. Wells. Toposes,triples and theories. Grundlehren der mathematischen Wissenschaften 278 , Springer-Verlag 1985 .

[2] B.S. Duran and P.L. Odell. Cluster anaysis. Lecture Notes in Economics and Mathematical Systems 100 , Springer 1974 .

[3] M.P. Fourman and D.S. Scott. Sheaves and logic.In "Applications of Sheaves" ,Lecture Notes in Mathematics 753 (1979) , 302 - 401 .

[4] U. Höhle. Quotients with respect to similarity relations.Fuzzy Sets and Systems 27 (1988), 31-44 .

[5] U. Höhle. Categorical foundations of cluster analysis. Preprint , Wuppertal 1988 .

[6] N. Jardine and R. Sibson. Mathematical taxonomy. Wiley 1971 .

[7] K. Menger. Geometry and positivism,a probabilistic microgeometry. In "Selected papers in logic and foundations, didactics, economics" Reidel 1979 .

[8] H. Poincaré. La science et l'hypothèse. Flammarion, Paris 1902 .

[9] H. Poincaré. La valeur de la science. Flammarion, Paris 1904 .

[10] B. Schweizer and A. Sklar. Probabilistic metric spaces. North-Holland, Amsterdam 1983 .

An Incremental Error Correcting Evaluation Algorithm

For Recursion Networks without Circuits

Ingo Althöfer
Fakultät für Mathematik, Universität Bielefeld
Postfach 8640, D-4800 Bielefeld 1
West Germany

Abstract

A recursion network consists of a directed graph together with an evaluation function v on the set of nodes of the graph, such that the value of every nonterminal node recursively results from the values of its successors. An example is a game graph with a grundy-function. Consider the problem to determine the value of a specified node 'root', when instead of v only an erroneous estimation function $\hat{v}$ on the set of all nodes is given. (Typically v and $\hat{v}$ will coincide in many, but not in all nodes.)

This paper presents a simple algorithm which yields the correct value $v(\text{root})$, if on every longpath with starting node 'root' the estimates $\hat{v}$ provide "more correct than wrong information" about the v-values. A longpath is a path that cannot be prolonged.

1. Introduction

A finite <u>recursion network</u> is a tuple

$R = (X, U, (N_x)_{x \in X}, (f_x)_{x \in X-L}, v)$, where $D = (X, U)$ is a finite <u>directed graph</u> with set of nodes X and set of edges $U \subset X^2$. $N_x \neq \emptyset$ is the set of all possible values for node $x \in X$, $f_x : (N_{y_1} \times N_{y_2} \times \ldots \times N_{y_{deg(x)}}) \to N_x$ is the recursion function at the nonterminal node $x \in X-L$, where $y_1, y_2, \ldots, y_{deg(x)}$ are the successors of x. L is the set of all terminal nodes in D. Finally the (admissible) evaluation function $v : X \to N := \bigcup_{x \in X} N_x$ satisfies

  (i) $v(x) \in N_x$ for all $x \in X$,

and

  (ii) $v(x) = f_x(v(y_1), \ldots, v(y_{deg(x)}))$ for all $x \in X-L$.

In this paper all recursion networks are finite. The bivalued game tree in Figure 1.1 is an example of a recursion network: The directed graph is a rooted tree, $N_x = N = \{-1, +1\}$ for all nodes $x \in X$, and $v(x) = f(\Gamma(x)) := -\min_{y \in \Gamma(x)} \{v(y)\}$ for all nodes $x \in X-L$, where $\Gamma(x)$ denotes the set of all successor nodes of x. In the figure nodes with value -1 (the LOSS-nodes) are labeled $\ominus$, nodes with value $+1$ (the WIN-nodes) are labeled $\oplus$.

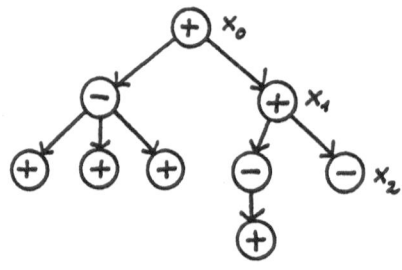

Figure 1.1

Another wellknown example of a recursion network is given by
$X = \{0,1,\ldots,n\}$, $U = \{(x,y) \in X^2 \mid (x-y) \in \{1,\ldots,t\} \text{ and } x \geq t\}$,
$N_x = N \neq \emptyset$ for all $x \in X$, and $f_x = f:N^t \to N$ for all
$x \in \{t,\ldots,n\}$. In this recursion scheme every "starting
tuple" $(v(0),\ldots,v(t-1)) \in N^t$ of length t determines
exactly one admissible evaluation function $v : X \to N$. More
generally, if $D=(X,U)$ is a graph without circuits, then
every evaluation of the terminal nodes yields exactly one
admissible evaluation function on X .

Given a finite recursion network, where the graph $D = (X,U)$
contains no circuits, and a specified node $r \in X$, this
article investigates the following question: How can we
determine $v(r)$, when our only information is an erroneous
estimation function $\hat{v} : X_r \to N$ of v with $\hat{v}(x) \in N_x$ for
all $x \in X_r$, where $X_r = \{x \in X \mid$ there exists a directed
path from r to x in $D\}$. Typically the evaluation function v and the v-estimation function $\hat{v}$ will coincide in
many, but not in all nodes x .

The result of this paper is an incremental algorithm which
runs in linear time (relative to the size of D) and yields
the correct value $v(r)$, if on every longpath with starting
node r the estimations $\hat{v}$ provide more correct than wrong
information about the v-values. A longpath is a path that
cannot be prolonged. In Figure 1.1 x_0-x_1-x_2 is a longpath,
whereas x_0-x_1 is not a longpath. An exact quantitative model
which is required for to speak of "more correct than wrong information" is provided in section 3 which also contains the
general algorithm and the result about its worst case perform-
ance. In section 2 we exhibit the algorithm by a concrete example. In section 4 we apply our (combinatorial) result to a

stochastic model in which the estimation errors $(\hat{v} \neq v)$ at different nodes occur independently of each other. Section 5 contains miscellaneous ideas.

2. An Example

Let $D = (X,U)$ be a binary rooted tree, $N_x = N = \{0,1\}$ for all $x \in X$, and $f : N^2 \to N$ defined by $f(n_1,n_2) \equiv n_1+n_2 \pmod 2$ for all $n_1,n_2 \in N$. Figure 2.1 shows such a tree with an admissible evaluation function v.

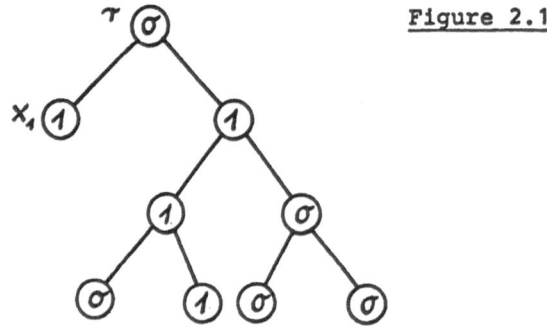

Figure 2.1

Given an erroneous estimation function $\hat{v} : X \to N$, the incremental algorithm works by recursively computing two functions $\tilde{v}_1 : X \to N$ and $\tilde{v}_2 : X \to N$, defined by

$$
\tilde{v}_1(x) := \begin{cases}
\hat{v}(x) , & \text{if } x \in L \\
f(\tilde{v}_1(y_1),\tilde{v}_1(y_2)) , & \text{if } x \in X-L \text{ and} \\
& \min\{\tilde{v}_2(y_1),\tilde{v}_2(y_2)\} \geq 1 \\
\hat{v}(x) , & \text{if } x \in X-L \text{ and} \\
& \min\{\tilde{v}_2(y_1),\tilde{v}_2(y_2)\} = 0 ,
\end{cases}
$$

and

$$
\tilde{v}_2(x) := \begin{cases}
1, & \text{if } x \in L \\
\min\{\tilde{v}_2(y_1), \tilde{v}_2(y_2)\}+1, & \text{if } x \in X-L \text{ and} \\
& \hat{v}(x) = f(\tilde{v}_1(y_1), \tilde{v}_1(y_2)) \\
|\min\{\tilde{v}_2(y_1), \tilde{v}_2(y_2)\}-1|, & \text{if } x \in X-L \text{ and} \\
& \hat{v}(x) \neq f(\tilde{v}_1(y_1), \tilde{v}_1(y_2)) \, .
\end{cases}
$$

$\tilde{v}_1(x)$ is an incremented estimate of the correct value $v(x)$.
$\tilde{v}_2(x)$ quantifies the strength of the estimate $\tilde{v}_1(x)$. The
next remark is a corollary of Remark 3.1 and Theorem 3.2 which
are presented and proved in section 3.

Remark 2.1: If in the tree D every longpath P with starting node r satisfies

$$\#\{x \in P \mid \hat{v}(x) = v(x)\} - \#\{x \in P \mid \hat{v}(x) \neq v(x)\} > 0, \qquad (2.1)$$

then $\tilde{v}_1(r) = v(r)$.

The incremental algorithm and Remark 2.1 are in some sense best
possible. Consider the tree of Figure 2.1 and another admissible
evaluation function

$$
v'(x) := \begin{cases}
1 - v(x), & \text{if } x \in \{r, x_1\} \\
v(x), & \text{otherwise,}
\end{cases}
$$

which differs from v in only one terminal node. The erroneous
estimation function $\hat{v}$, defined by

$$
\hat{v}(x) = \begin{cases}
0, & \text{if } x \in \{r, x_1\} \\
v(x), & \text{otherwise}
\end{cases}
$$

differs from v as well as from v' in only one node. Furthermore it satisfies the inequality (2.1) with both evaluation functions v and v' for all longpaths P except

$r-x_1$, for which equality holds.

Similar examples can be constructed in almost all nondegenerate recursion networks.

3. Weighted Information and the General Result

Given a recursion network without circuits and a nonnegative weighting function $w : X \to \mathbb{R}_{\geq 0}$, we define and compute recursive functions $\tilde{v}_1 : X \to N$ and $\tilde{v}_2 : X \to \mathbb{R}_{\geq 0}$. For simplicity we use the notion $\tilde{v}(\bar{y})$ instead of

$$(\tilde{v}(y_1), \ldots, \tilde{v}(y_{\deg(x)})) .$$

$$\tilde{v}_1(x) := \begin{cases} \hat{v}(x) , & \text{if } x \in L \\ f(\tilde{v}_1(\bar{y})) , & \text{if } x \in X-L \text{ and} \\ \quad \min_{y \in \Gamma(x)} \{\tilde{v}_2(y)\} \geq w(x) \\ \hat{v}(x) , & \text{if } x \in X-L \text{ and} \\ \quad \min_{y \in \Gamma(x)} \{\tilde{v}_2(y)\} < w(x) . \end{cases}$$

$$\tilde{v}_2(x) := \begin{cases} w(x) , & \text{if } x \in L \\ \min_{y \in \Gamma(x)} \{\tilde{v}_2(y)\} + w(x) , & \text{if } x \in X-L \text{ and} \\ \quad \hat{v}(x) = f(\tilde{v}_1(\bar{y})) \\ \left| \min_{y \in \Gamma(x)} \{\tilde{v}_2(y)\} - w(x) \right| , & \text{if } x \in X-L \text{ and} \\ \quad \hat{v}(x) \neq f(\tilde{v}_1(\bar{y})) \end{cases}$$

The weighting $w(x)$ quantifies how reliable the information $\hat{v}(x)$ is supposed to be. In the example in section 2 we had simply chosen $w(x) = 1$ for all $x \in X$.

We define $L_0 := L$ and

$$L_{t+1} := \{x \in (X- \bigcup_{i=0}^{t} L_i) \mid \Gamma(x) \subset \bigcup_{i=0}^{t} L_i\} \text{ for all } t \in N .$$

$\bigcup_{t=0}^{\infty} L_t = X$, if the finite graph D contains non circuits.

First of all $\tilde{v}_1$ and $\tilde{v}_2$ are computed for the nodes in L_0, then for the nodes in L_1, and so on.

Given a recursion network without circuits and an estimation function $\hat{v}$, the following function $d : X \to \mathbb{R}$ is well-defined.

$$d(x) := \begin{cases} + w(x), & \text{if } x \in L \text{ and } v(x) = \hat{v}(x) \\ - w(x), & \text{if } x \in L \text{ and } v(x) \neq \hat{v}(x) \\ \min_{y \in \Gamma(x)} \{d(y)\} + w(x), & \text{if } x \in X-L \text{ and } v(x) = \hat{v}(x) \\ \min_{y \in \Gamma(x)} \{d(y)\} - w(x), & \text{if } x \in X-L \text{ and } v(x) \neq \hat{v}(x). \end{cases}$$

Remark 3.1:

$$d(x) = \min \left\{ \sum_{\substack{z \in P \\ \text{and} \\ v(z) = \hat{v}(z)}} w(z) - \sum_{\substack{z \in P \\ \text{and} \\ v(z) \neq \hat{v}(z)}} w(z) \;\middle|\; \begin{array}{l} P \text{ is longpath with} \\ \text{starting node } x \end{array} \right\}.$$

Hence $d(x)$ is positive, iff every longpath with starting node x contains more correct than wrong information $\hat{v}$ about v.

Theorem 3.2: If a recursion network without circuits and an estimation function $\hat{v}$ are given, then the functions $d, \tilde{v}_1$, and $\tilde{v}_2$ satisfy for every node $x \in X$:

(i) $\tilde{v}_1(x) = v(x)$ and $\tilde{v}_2(x) \geq d(x)$, if $d(x) > 0$

and

(ii) $\tilde{v}_1(x) = v(x)$ or $\tilde{v}_2(x) \leq -d(x)$, if $d(x) \leq 0$.

Proof: Proceeds by induction on L_t. Clearly the case $t = 0$ holds. Now assume $x \in L_t$, $t \geq 1$. We define $d^*(x) = \min_{y \in \Gamma(x)} \{d(y)\}$. The further proof distinguishes six cases.

Case (α): $\quad$ $d(x) > 0$ and $d*(x) > 0$.

$\quad$ Thus by induction $\tilde{v}_1(y) = v(y)$ for all $y \in \Gamma(x)$,

$\quad$ $f(\tilde{v}_1(\bar{y})) = v(x)$, and $w(x) < d*(x)$, if $\hat{v}(x) \neq v(x)$.

$\quad$ Hence $\tilde{v}_1(x) = v(x)$ and

$\quad$ $\tilde{v}_2(x) = \min\limits_{y \in \Gamma(x)} \{\tilde{v}_2(y)\} \pm w(x) \geq d*(x) \pm w(x) = d(x)$.

Case (β): $\quad$ $d(x) > 0$ and $d*(x) \leq 0$

$\quad$ Thus $\hat{v}(x) = v(x)$, $w(x) \geq d(x)$, and $w(x) + d*(x) = d(x)$

$\quad$ Subcase: $\quad$ $f(\tilde{v}_1(\bar{y})) = v(x)$.

$\quad\quad$ Hence $\tilde{v}_1(x) = v(x)$ and $\tilde{v}_2(x) \geq w(x) \geq d(x)$.

$\quad$ Subcase: $\quad$ $f(\tilde{v}_1(\bar{y})) \neq v(x)$.

$\quad\quad$ Hence by induction $\min\limits_{y \in \Gamma(x)} \{\tilde{v}_2(y)\} \leq -d*(x)$,

$\quad\quad$ $\tilde{v}_1(x) = \hat{v}(x)$, and

$\quad\quad$ $\tilde{v}_2(x) \geq w(x) - \min\limits_{y \in \Gamma(x)} \{\tilde{v}_2(y)\} \geq w(x) + d*(x) = d(x)$.

Case (γ): $\quad$ $d(x) \leq 0$ and $\hat{v}(x) = v(x)$.

$\quad$ Thus $d*(x) + w(x) = d(x)$ and $d*(x) \leq d(x) \leq 0$.

$\quad$ Subcase: $\quad$ $f(\tilde{v}_1(\bar{y})) = v(x)$.

$\quad\quad$ Hence $\tilde{v}_1(x) = v(x)$.

$\quad$ Subcase: $\quad$ $f(\tilde{v}_1(\bar{y})) \neq v(x)$.

$\quad\quad$ Hence by induction $\min\limits_{y \in \Gamma(x)} \{\tilde{v}_2(y)\} \leq -d*(x)$

$\quad\quad$ and $\tilde{v}_2(x) \leq -d*(x) - w(x) = -d(x)$, if $\tilde{v}_1(x) \neq v(x)$.

Case (δ): $\quad$ $d(x) \leq 0$, $\hat{v}(x) \neq v(x)$, and $d*(x) > 0$.

$\quad$ Thus $d(x) = d*(x) - w(x)$, and by induction $f(\tilde{v}_1(\bar{y})) = v(x)$

$\quad$ and $\min\limits_{y \in \Gamma(x)} \{\tilde{v}_2(y)\} \geq d*(x)$.

$\quad$ Subcase: $\quad$ $\tilde{v}_1(x) = v(x)$. Ready.

Subcase: $\tilde{v}_1(x) \neq v(x)$.

Hence $\tilde{v}_2(x) = w(x) - \min_{y \in \Gamma(x)} \{\tilde{v}_2(y)\}$

$\leq w(x) - d*(x) = -d(x)$.

Case (ε): $d(x) \leq 0$, $\hat{v}(x) \neq v(x)$, $d*(x) \leq 0$, and $\tilde{v}_1(x) \neq v(x)$.

Thus $d(x) = d*(x) - w(x)$.

Subcase: $f(\tilde{v}_1(\bar{y})) = v(x)$.

Hence $\min_{y \in \Gamma(x)} \{\tilde{v}_2(y)\} < w(x)$ and

$\tilde{v}_2(x) \leq w(x) = d*(x) - d(x) \leq -d(x)$.

Subcase: $f(\tilde{v}_1(\bar{y})) \neq v(x)$.

Hence $\min_{y \in \Gamma(x)} \{\tilde{v}_2(y)\} \leq -d*(x)$ and

$\tilde{v}_2(x) \leq -d*(x) + w(x) = -d(x)$.

Case (φ): $d(x) \leq 0$, $\hat{v}(x) \neq v(x)$, $d*(x) \leq 0$, and $\tilde{v}_1(x) = v(x)$.

Ready.

■

4. A Stochastic Model with Independent Errors

In [1] we had investigated the bivalued game tree model, i.e.
the recursion network consisted of a rooted tree $T = (X,U)$,
$N = \{-1,+1\}$, and $v(x) = -\min_{y \in \Gamma(x)} \{v(y)\}$ for all nonterminal
nodes x. Given an erroneous estimation function $\hat{v}: X \to \{-1,+1\}$,
we had computed a function $\tilde{v} : X \to \mathbb{Z}$ recursively by

$$\tilde{v}(x) = \begin{cases} \hat{v}(x), & \text{if } x \in L \\ -\min_{y \in \Gamma(x)} \{\tilde{v}(y)\} + \hat{v}(x), & \text{if } x \in X-L \end{cases}$$

to get a "better" estimate

$$v^*(\text{root}) := \begin{cases} +1, & \text{if } \tilde{v}(\text{root}) \geq 0 \\ -1, & \text{if } \tilde{v}(\text{root}) < 0 \end{cases}$$

for the root value than $\hat{v}(\text{root})$.

Under the assumptions that all longpaths of the tree have length t , that every node has at most k successors, and that estimation errors $\hat{v} \neq v$ at different nodes occur stochastically and independently of each other we had proved

Theorem 4.1: There exist constants c_k, $0 < c_k < 1$, and ε_k, $0 < \varepsilon_k < 1$, such that

$$\text{Prob } \{v^*(\text{root}) \neq v(\text{root})\} \leq c_k^t ,$$

if Prob $\{\hat{v}(x) \neq v(x)\} \leq \varepsilon_k$ for all nodes $x \in X$.

A comparable result can be obtained for general recursion networks without circuits, when $\max_{x \in X} \deg(x) \leq k$, $w(x)=1$ for all $x \in X$, and all longpaths with starting node r have the same length t . If on every longpath the estimation errors at different nodes of the path occur stochastically and independently of each other, we get

Theorem 4.2: There exist constants c_k, $0 < c_k < 1$, and ε_k, $0 < \varepsilon_k < 1$, such that the incremental algorithm of section 3 yields

$$\text{Prob } \{\tilde{v}_1(r) \neq v(r)\} \leq c_k^t ,$$

if Prob $\{\hat{v}(x) \neq v(x)\} \leq \varepsilon_k$ for all $x \in X$.

Proof: Let $P = r-x_1-\ldots-x_t$ be a longpath in the recursion network. P is called <u>incorrect</u>, if $\hat{v}(x) \neq v(x)$ for at least $\lceil \frac{t+1}{2} \rceil$ many nodes $x \in P$.

$$\text{Prob } \{P \text{ is incorrect}\} \leq \sum_{i=\lceil\frac{t+1}{2}\rceil}^{t+1} \binom{t+1}{i} \epsilon_k^i \leq \sum_{i=\lceil\frac{t+1}{2}\rceil}^{t+1} 2^{t+1} \epsilon_k^{\lceil\frac{t+1}{2}\rceil}$$

$$\leq \left(t+2-\lceil\tfrac{t+1}{2}\rceil\right)\left(4\ \epsilon_k\right)^{\lceil\frac{t+1}{2}\rceil}.$$

With the notion

$E_{error} :=$ expected number of incorrect longpaths with starting node r

we get

$$\text{Prob } \{\tilde{v}_1(r) \neq v(r)\} \leq E_{error}$$

$$= \sum_{P \text{ longpath}} \text{Prob } \{P \text{ is incorrect}\}$$

$$\leq k^t (t+2-\lceil\tfrac{t+1}{2}\rceil)(4\ \epsilon_k)^{\lceil\frac{t+1}{2}\rceil}$$

$$\leq (t+2-\lceil\tfrac{t+1}{2}\rceil)(4k^2\epsilon_k)^{\lceil\frac{t+1}{2}\rceil}.$$

Hence the theorem holds for every $0 < \epsilon_k < \dfrac{1}{4k^2}$.

∎

5. Miscellaneous

(a) In the introduction we had mentioned as an example the recursion $v(m) = f(v(m-1),\ldots,v(m-t))$ for $m = t,\ldots,n$.

Remark 5.1: If $\#\{m \in \{0,\ldots,n\} \mid \hat{v}(m) \neq v(m)\} \leq \dfrac{n-t+1}{2t}$ and $w(m) = 1$ for all $m \in \{0,\ldots,n\}$, then the incremental algorithm yields $\tilde{v}_1(n) = v(n)$.

Proof: The condition of the remark implies that every longpath in the underlying graph contains more nodes with correct $\hat{v}$-value than nodes with wrong $\hat{v}$-value. Hence Theorem 3.2 can be applied.

∎

(b) An incremental algorithm similar to that of section 3 can also be obtained for finite recursion networks <u>with</u> circuits, if besides the condition for finite longpaths every infinite path $(r = x_0) - x_1 - x_2 - \ldots$ satisfies

$$\lim_{k \to \infty} \inf \left[\sum_{\substack{0 \le i \le k \text{ and} \\ \hat{v}(x_i) = v(x_i)}} w(x_i) - \sum_{\substack{0 \le i \le k \text{ and} \\ \hat{v}(x_i) \ne v(x_i)}} w(x_i) \right] > 0 \, .$$

(c) In certain recursion networks the condition "<u>Every</u> longpath contains more correct than wrong information" of Theorem 3.2 can be weakened. For example the incremental negamax algorithm $\tilde{v}$ in section 4 already yields the correct rootvalue if all longpaths of one <u>solution tree</u> contain more correct than wrong information. (The solution tree [5] is only a "small" subtree of the complete game tree. In a tree of depth t with regular branching degree k a solution tree contains only $k^{\lceil \frac{t}{2} \rceil}$ or $k^{\lfloor \frac{t}{2} \rfloor}$ of the k^t leaf nodes, depending on whether the root value is -1 or $+1$, respectively.)

(d) The incremental algorithm works also in the presence of arbitrary real-valued weightings $w(x)$. Theorem 3.2, however, holds only when all weightings are nonnegative. The reason for this limitation is that negative weightings (only) say that the information $\hat{v}(x) \in N_x$ is supposed to be wrong. This is for sets N_x with more than two elements a weaker information than any with $w(x) > 0$.

References

1. I. Althöfer, An incremental negamax algorithm, 1987.

2. R. Bellman, S. Dreyfus, "Applied Dynamic Programming",
   Princeton University Press, Princeton (New Jersey), 1962.

3. J. H. van Lint, "Introduction to Coding Theory",
   Springer, New York, 1982.

4. N.J. Nilsson, "Principles of Artificial Intelligence",
   Tioga, Palo Alto (California), 1980.

5. J. Pearl, "Heuristics-Intelligent Search Strategies for
   Computer Problem Solving", Addison-Wesley, Reading
   (Massachusetts), 1985.

6. J. Pearl, Distributed revision of composite beliefs,
   Artificial Intelligence 33 (1987), 173-215.

7. W.W. Peterson, E.J. Weldon Jr., "Error Correcting Codes",
   2. ed., MIT Press, Cambridge (Massachusetts), 1972.

Volume 18

S. A. Levin, Cornell University, Ithaca, NY; T. G. Hallam, L. J. Gross, University of Tennessee, Knoxville, TN (Eds.)

Applied Mathematical Ecology

1989. XIV, 491 pp. 114 figs. Hardcover DM 98,– ISBN 3-540-19465-7

Contents: Introduction. – Resource Management. – Epidemiology: Fundamental Aspects of Epidemiology Case Studies. – Ecotoxicology. – Demography and Population Biology. – Author Index. – Subject Index.

This book builds on the basic framework developed in the earlier volume – "Mathematical Ecology", edited by T. G. Hallam and S. A. Levin, Springer 1986, which lays out the essentials of the subject. In the present book, the applications of mathematical ecology in ecotoxicology, in resource management, and epidemiology are illustrated in detail. The most important features are the case studies, and the interrelatedness of theory and application. There is no comparable text in the literature so far. The reader of the two-volume set will gain an appreciation of the broad scope of mathematical ecology.

Volume 19

J. D. Murray, Oxford University

Mathematical Biology

1989. XIV, 767 pp. 292 figs. Hardcover DM 98,– ISBN 3-540-19460-6

This textbook gives an in-depth account of the practical use of mathematical modelling in several important and diverse areas in the biomedical sciences.
The emphasis is on what is required to solve the real biological problem. The subject matter is drawn, for example, from population biology, reaction kinetics, biological oscillators and switches, Belousov-Zhabotinskii reaction, neural models, spread of epidemics.
The aim of the book is to provide a thorough training in practical mathematical biology and to show how exciting and novel mathematical challenges arise from a genuine interdisciplinary involvement with the biosciences. It also aims to show how mathematics can contribute to biology and how physical scientists must get involved.
The book also presents a broad view of the field of theoretical and mathematical biology and is a good starting place from which to start genuine interdisciplinary research.

In preparation

Volume 20

J. E. Cohen, Rockefeller University, New York, NY; F. Briand, Gland, Switzerland; C. M. Newman, University of Arizona, Tucson, AZ

Community Food Webs

Data and Theory

1990. Approx. 300 pp. 46 figs. ISBN 3-540-51129-6

Springer-Verlag Berlin
Heidelberg New York London
Paris Tokyo Hong Kong

Springer

Journal of Mathematical Biology

For mathematicians and biologists working in a wide variety of fields – genetics, demography, ecology, neurobiology, epidemiology, morphogenesis, cell biology – the **Journal of Mathematical Biology** publishes:

- papers in which mathematics is used for a better understanding of biological phenomena
- mathematical papers inspired by biological research, and
- papers which yield new experimental data bearing on mathematical models.

Editorial Board: K. P. Hadeler, Tübingen; S. A. Levin, Ithaca (Managing Editors); H. T. Banks, Los Angeles; J. D. Cowan, Chicago; J. Gani, Santa Barbara; F. C. Hoppensteadt, East Lansing; D. Ludwig, Vancouver; J. D. Murray, Oxford; T. Nagylaki, Chicago; L. A. Segel, Rehovot

Subscription Information:
ISSN 0303-6812 Titel No. 285
1990, Vol. 28 (6 issues)
DM 712,- plus carriage charges (FRG DM 11,88; other countries DM 17,40)

From the contents:

Springer

Lecture Notes in Biomathematics